PRECOLLEGE SCIENCE AND MATHEMATICS TEACHERS

Monitoring Supply, Demand, and Quality

Dorothy M. Gilford and Ellen Tenenbaum, Editors

Panel on Statistics on Supply and Demand for Precollege
Science and Mathematics Teachers
F. Thomas Juster, Chair

Committee on National Statistics
Commission on Behavioral and Social Sciences and Education
National Research Council

NATIONAL ACADEMY PRESS
Washington, D.C. 1990

NOTICE: The project that is the subject of this report was approved by the Governing Board of the National Research Council, whose members are drawn from the councils of the National Academy of Sciences, the National Academy of Engineering, and the Institute of Medicine. The members of the committee responsible for the report were chosen for their special competences and with regard for appropriate balance.

This report has been reviewed by a group other than the authors according to procedures approved by a Report Review Committee consisting of members of the National Academy of Sciences, the National Academy of Engineering, and the Institute of Medicine.

The National Academy of Sciences is a private, nonprofit, self-perpetuating society of distinguished scholars engaged in scientific and engineering research, dedicated to the furtherance of science and technology and to their use for the general welfare. Upon the authority of the charter granted to it by the Congress in 1863, the Academy has a mandate that requires it to advise the federal government on scientific and technical matters. Dr. Frank Press is president of the National Academy of Sciences.

The National Academy of Engineering was established in 1964, under the charter of the National Academy of Sciences, as a parallel organization of outstanding engineers. It is autonomous in its administration and in the selection of its members, sharing with the National Academy of Sciences the responsibility for advising the federal government. The National Academy of Engineering also sponsors engineering programs aimed at meeting national needs, encourages education and research, and recognizes the superior achievements of engineers. Dr. Robert M. White is president of the National Academy of Engineering.

The Institute of Medicine was established in 1970 by the National Academy of Sciences to secure the services of eminent members of appropriate professions in the examination of policy matters pertaining to the health of the public. The Institute acts under the responsibility given to the National Academy of Sciences by its congressional charter to be an adviser to the federal government and, upon its own initiative, to identify issues of medical care, research, and education. Dr. Samuel O. Thier is president of the Institute of Medicine.

The National Research Council was organized by the National Academy of Sciences in 1916 to associate the broad community of science and technology with the Academy's purposes of furthering knowledge and advising the federal government. Functioning in accordance with general policies determined by the Academy, the Council has become the principal operating agency of both the National Academy of Sciences and the National Academy of Engineering in providing services to the government, the public, and the scientific and engineering communities. The Council is administered jointly by both Academies and the Institute of Medicine. Dr. Frank Press and Dr. Robert M. White are chairman and vice chairman, respectively, of the National Research Council.

This project was supported with funds from the National Science Foundation and the National Center for Education Statistics, U.S. Department of Education.

Library of Congress Catalog Card No. 89-89-64263
International Standard Book Number 0-309-04197-X

Additional copies of this report are available from:
National Academy Press
2101 Constitution Avenue, NW
Washington, DC 20418

S099

Printed in the United States of America

PANEL ON STATISTICS ON SUPPLY AND DEMAND FOR PRECOLLEGE SCIENCE AND MATHEMATICS TEACHERS

F. THOMAS JUSTER (Chair), Institute for Social Research, University of Michigan
WILMER S. CODY, State Department of Education, Baton Rouge, Louisiana
GLENN A. CROSBY, Department of Chemistry, Washington State University
F. JOE CROSSWHITE, Department of Mathematics, Northern Arizona University
HARRIET FISHLOW, Undergraduate Enrollment Planning, Office of the President, Academic Affairs Division, University of California, Berkeley
CHARLOTTE V. KUH, Graduate Record Examination Board, Princeton, New Jersey
EUGENE P. MCLOONE, Department of Education Policy, Planning, and Administration, University of Maryland
MICHAEL S. MCPHERSON, Department of Economics, Williams College
RICHARD J. MURNANE, Graduate School of Education, Harvard University
INGRAM OLKIN, Department of Statistics and School of Education, Stanford University
JOHN J. STIGLMEIER, Information Center on Education, New York State Education Department

DOROTHY M. GILFORD, Study Director
ELLEN TENENBAUM, Consultant
M. JANE PHILLIPS, Administrative Secretary

COMMITTEE ON NATIONAL STATISTICS

Acknowledgments

The panel wishes to thank the many people who contributed to the development of this report. First, we benefited greatly from the experiences of those school district officials and teachers who participated in the case studies. Special thanks are extended to consultants Jane L. David, Marianne Amarel, and Jennifer P. Pruyn, who conducted in-depth case studies of six of the selected school districts. Also of great benefit to the panel were the insights shared by the personnel directors of the seven large city school systems during the conference that took place in May 1988: Charles Almo, Chicago Public Schools; Edward Aquilone, New York City Schools; Ray Cohrs, Seattle Public Schools; Thomas Killeen, Los Angeles Unified School System; Alan Olkes, Dade County School District; George Russell, San Diego City Unified School District; and Jim Shinn, Montgomery County School System. Marlene Holayter, Fairfax County Public Schools, helped in organizing the conference by identifying potential participants; she also participated in the conference.

Particular thanks are due to Richard Berry and to Ronald Anderson of the Office of Studies and Program Assessment, the National Science Foundation. Each of them served as the panel's project officer in different phases of the panel, provided valuable information to panel members, and shared their thoughts and professional experience. The panel is also grateful for the help and encouragement extended by Paul Planchon of the National Center for Education Statistics.

We would also like to thank members of the Commission on Behavioral and Social Sciences and Education and of the Committee on National Statistics, who reviewed the report and provided thoughtful and incisive comments. In addition, Mary Papageorgiou, statistician with the National Center for Education Statistics (NCES), carefully read the draft report and clarified important information about the Schools and Staffing Survey now being undertaken at NCES. Constance Citro of the Committee on National Statistics also found time to review the report and offered cogent comments, reflecting her earlier service to the panel as study director for the first phase of the project. And Jane L. David took the time to read the draft report, raising important questions that led to rewriting and reorganizing parts of the report.

Christine McShane, editor of the Commission on Behavioral and Social Sciences and Education, provided highly professional editing of the report, which contributed greatly to its readability. She also prepared it for final publication.

The panel extends its appreciation to its staff for their tireless work and dedication to this project. Jane Phillips served ably as administrative secretary for the panel, taking care of the logistical arrangements for the panel meetings and cheerfully and efficiently handling the numerous rounds of revisions to the draft report. Dorothy M. Gilford, who served as study director for the second phase, not only contributed substantially to the style, substance, and coherence of the report, but also successfully motivated panel members to work much harder on the project than they may have originally planned. Ellen Tenenbaum assisted Dorothy in producing a cohesive synthesis of our discussions and conclusions, as well as in rewriting and reorganizing drafts prepared by panel members.

Finally, I wish to thank the panel members themselves for their generous contributions of time and expert knowledge. It has been a pleasure to work together toward this final report. While the findings and recommendations reflect the collective thinking of the panel, I appreciate the hard work done by individual panel members in drafting text for several chapters: Harriet Fishlow prepared the demography section of Chapter 2; Richard Murnane drafted Chapter 3; John Stiglmeier prepared the introduction to Chapter 4 and provided very useful data about teachers in the state of New York; Eugene McLoone drafted most of Chapter 5; and Charlotte Kuh wrote the introduction to Chapter 6. It would be remiss not to mention Dick Murnane's and Charlotte Kuh's helpful responses to many questions raised by the staff while the report was being revised to respond to reviewers.

Thomas Juster, Chair
Panel on Statistics on Supply and Demand
for Precollege Science and Mathematics Teachers

Contents

Summary

The activities of an increasingly technological society call for greater command of science and mathematics at the precollege level than at any time in the recent past. Yet evidence from numerous studies indicates that the majority of American students are not being equipped with the scientific and mathematical tools needed to participate in that technological society. International comparisons of secondary students' achievement show Americans generally ranking in the lowest half of the distribution among their counterparts in developed countries. There has also been increasing concern about shortages of qualified individuals who will teach science and mathematics at the elementary or secondary levels. Studies point to an increased demand for science and mathematics teachers and as well as an inadequate supply of highly qualified individuals to fill those positions.

To what extent is there a shortage, in terms of both quantity and quality, of science and mathematics teachers? National data and research related to teacher supply and demand in science and mathematics are insufficient to substantiate statistical conclusions or to prescribe specific national policies.

Concerns about shortages in these critical areas and about the quality of the available statistics led the National Science Foundation and the National Center for Education Statistics of the U.S. Department of Education to ask the National Research Council to evaluate the statistics on the supply and demand for science and mathematics teachers.

In 1986 the National Research Council established the Panel on Statistics on Supply and Demand for Precollege Science and Mathematics Teachers to conduct such a study, in two phases. In the first phase the panel reviewed teacher supply and demand models in selected states and the national model used by the National Center for Education Statistics. The tentative conclusion of the panel was that none of the available models or data were adequate to assess either shortage or quality. The principal difficulty, as seen by the panel, was on the supply side of the models, although there were also serious shortcomings in important aspects of the demand side. We recommended in the interim report (1987c:7) that research on teacher supply be conducted, foremost on the behavioral determinants of key groups of new entrants—new graduates, former teachers, and persons hired on emergency certification. The forces underlying teacher migration were a second issue of behavioral research. Also in need of study were the behavioral and environmental factors influencing attrition. The panel also identified information on teacher qualifications that could be collected for use in descriptive profiles and in supply and demand models. The first phase culminated in the 1987 interim report, *Toward Understanding Teacher Supply and Demand: Priorities for Research and Development* (National Research Council, 1987c).

In its second phase of work the panel continued with a more detailed investigation of statistics and models of the supply and demand for science and mathematics teachers. Case studies were conducted in 39 school districts to identify variables that might usefully be included in such statistics. The statistical basis for describing entry and exit patterns of science and mathematics teachers was examined. A conference of personnel directors of seven large school systems was held in May 1988 to discuss data available relating to the supply, demand, and qualifications of science and mathematics teachers. The panel also conducted a comprehensive review of state statistical data pertinent to teacher supply and demand models and the availability of such data to researchers.

As indicated in the panel's final report, there is great diversity in labor market situations and in the actions taken by applicants and school systems to balance supply and demand. Some of these important interactions, while not easily portrayed statistically, are essential to understanding the nation's supply and demand situation as it affects science and mathematics at the precollege level. The recommendations in Chapter 6 of the report reflect the insights gleaned throughout the second phase of the panel's work. In this report, we go beyond the panel's interim report and try to provide a more specific assessment both of the available data and of data that could potentially be obtained that would enrich existing models. We also try to provide better insights into the relationship between quality and issues relating to supply and demand.

In this report the panel has evaluated the statistics on supply, demand, and quality as they pertain to science and mathematics teachers. We conclude that available data on the supply aspects of teacher labor markets and on the quality-related adjustments that bring supply and demand into equilibrium remain inadequate, although we recognize that the National

Center for Education Statistics has taken major steps toward an improved national data base.

This summary presents the recommendations to have the highest priority. Most of the recommendations pertaining to improved data are addressed to the National Center for Education Statistics (NCES). Recommendations related to further research topics are addressed to the National Science Foundation (NSF) and the education research community at large. A recommendation for research facilitation is directed to the Department of Education's Office of Educational Research and Improvement, and our final recommendation calls for NCES to convene a series of conferences on issues of teacher supply, demand, and quality.

DATA RECOMMENDATIONS

Meaningful descriptions of supply and demand for precollege science and mathematics teachers, their interactions, and the role of quality in bringing supply and demand into balance require comprehensive national data. The Schools and Staffing Survey (SASS), first conducted in 1987-88 under the aegis of the National Center for Education Statistics, represents a groundbreaking effort to capture some of the most relevant data on a regular basis. If this survey is repeated periodically and disseminated quickly, it holds particular promise for providing statistics on a number of key aspects of supply, demand, and quality. As with all new surveys, the extent to which SASS will meet its goals cannot be known until the data have been received and analyzed.[1] The SASS data, used in conjunction with other NCES surveys and the panel's additional recommendations for data enhancement, should provide the basic data for monitoring the supply, demand, and quality of teachers and for preparing informative reports and analyses. However, current data collection efforts and our knowledge of the relation between incentives, quality, and supply are still inadequate to support meaningful behavioral models of teacher supply and demand. Thus, we recommend a sequential approach:

- First, as efforts are made to improve the consistency, scope, and quantity of data, publish indicators from existing data that are considered relevant to teacher supply, demand, and quality.

[1] The release of the SASS data base has been delayed a few months in order to guarantee the confidentiality of the data. The recent Hawkins-Stafford Act (P.L. 100-297) requires that the National Center for Education Statistics protect the privacy of individually identifiable information collected through survey questionnaires. NCES has recently formed a Disclosure Review Board, consisting of senior NCES staff and outside experts, to establish procedures and then review data products in order to release the maximum amount of data while protecting the privacy of survey respondents. The first release of SASS tables is scheduled for spring 1990.

- Second, carry out the research needed to support behavioral models.
- Third, as data bases are improved and research findings on the relation between incentives and supply become available, devote resources to structural modeling that goes beyond simple extrapolative projection.

The specific recommendations presented below address the need for resources for SASS; data related to teacher demand, supply and quality; and a group of desirable general data practices. The complete set of recommendations and their discussion appear in Chapter 6.

Resources for Data

In the near term it is essential to monitor the state of demand, supply, and quality of precollege science and mathematics teachers, and we advance four recommendations to that end. The NCES has recognized the need for a major effort to gain better information concerning teachers and has initiated SASS, which, if it lives up to its promise, has the potential to provide the best data on teachers this country has had. Analyses based on SASS can improve the nation's understanding of the supply, demand, and quality of precollege science and mathematics teachers.

Priority Recommendation 1. We recommend that provision be made in the budget for the National Center for Education Statistics to conduct the Survey of Schools and Staffing on a regular cycle and that the budget include funds for follow-up surveys of teachers who leave teaching and for in-house and external analysis of the survey data.

Demand Data

Estimates of the demand for hiring teachers depend on at least three components: student enrollment, pupil-teacher ratios, and teacher attrition rates. (It should be noted that teacher attrition is largely a supply phenomenon, reflecting the decisions of individual teachers. In Chapters 3 and 4, we treat attrition as a supply variable, but for some purposes it is a natural transformation to think of it as resulting in a demand for new teachers.) Although the task of projecting enrollment-driven demand for science and mathematics teachers is fairly straightforward, and most of the necessary data are available through the Bureau of the Census and NCES, the data on teacher attrition are deficient. At any organizational level, whether national, state, or district, attrition is generally defined as the number of teachers who taught in that organizational unit in one year but not in that unit in a similar position in the following year. The most recent NCES estimates of attrition rely on 1983 data from the Bureau of Labor Statistics and are not disaggregated by discipline. The data most needed now for better projecting teacher demand are attrition data, although needed improvements in other demand-related data are also noted. Data on attrition for reasons other than retirement are of particular interest.

Priority Recommendation 2. We recommend that NCES collect data on attrition rates classified by retirement or other cause, and by discipline, as a first priority. Other data recommendations are for annual information on state-mandated course requirements and periodic data on changes in science and mathematics course offerings and enrollments.

Supply Data

The major shortcomings of current supply-demand models and reports of teacher supply and demand occur on the supply side.

The primary components of supply are continuing teachers and new entrants. The supply of continuing teachers is estimated using attrition rates, which are much in need of improvement, as the preceding recommendation and the one below emphasize. Estimating the supply of new entrants is more difficult still, since most new entrants do not come directly from teacher training institutions. Although a few come from alternative certification programs, the major source of new entrants is the reserve pool, which consists of people with teaching experience who did not teach last year or individuals who were certified to teach at least a year ago but who have never taught. Data on these key sources are inadequate or nonexistent, though SASS is making significant inroads toward a national data base that will describe these components more clearly.

Effective monitoring of supply must include information on the quality of the supply (described below). To construct behavioral models of supply, data will be needed that capture behavioral aspects of supply. Our highest-priority data recommendations for the near term call for better data on newly certified individuals and their incentives to teach, on the components of the reserve pool, and on retention and attrition patterns. Priority Recommendation 3 is therefore widely encompassing, including not only components of quality, but also data that capture behavioral aspects of supply.

Priority Recommendation 3: We recommend that NCES collect the specified data (in order of priority under each of the three headings) on the following aspects of teacher supply.

(a) *New hires and incentives to teach*:
- *Comparative salary data to indicate competitiveness of teachers' salaries relative to those of alternative nonteaching positions.*
- *Data on reasons why teachers selected their current school/district.*

- *Data on the number of last year's certificants, by type of certificate, who were hired (or received a firm job offer) by school districts and the proportion of those who applied for positions who were hired.*
- *Trend data from districts on the ratio of the number of applicants to vacancies in teaching, by field, and on the number of job offers per vacancy.*

(b) *The supply potential of the major components of the reserve pool:*

- *Data following new college graduates over time, to determine the proportion that enter teaching by the number of years after graduation, reasons for leaving teaching, time spent out of teaching, and reentry into teaching.*
- *Retrospective data that track new hires from the reserve pool backward, to study their career histories prior to entering or reentering teaching.*
- *Data on those certified in a given state who are not currently teaching in that state.* Using data from state certification files, states can track certificants who still live in the state and characterize that segment of the reserve pool by age, subject, specialty, years of past teaching experience, and interest in teaching.

(c) *Retention and attrition rates:*

- *Data from schools on the distribution of teachers by age, race/ethnicity, sex, and disciplinary area, as well as attrition levels within these categories. Attrition should be classified by retirement or other cause.*
- *Information from former science and mathematics teachers on incentives to leave teaching.*

Quality Data

The notion of an adequate supply of science and mathematics teachers must be understood in terms of the quality of that supply. In the short term it is in large part through adjustments in quality that supply and demand come into equilibrium. In responding to perceived shortages, school systems may opt over the long term to increase salaries or improve working conditions. In the short term, they may recruit more aggressively, increase class sizes, or cancel courses. Frequently, the adjustment mechanism operates through changes in the quality of individuals hired. Hence, there may be no observed quantity imbalance but instead a change in the quality characteristics of the teaching force. It is critical to obtain statistics that relate to quality, but very little information exists that helps to define or measure quality at present. We need to know more about how quality can be measured and how quality adjustments take place. The next recommendations encompass a wide variety of data needed toward that end.

Priority Recommendation 4. To provide indicators of aspects of the quality of teachers and aspects of the school system that affect either teacher quality or teaching quality, we recommend that the data listed below (in order of priority under each of the two headings) be collected and monitored over time.

(a) Qualifications of teachers:

- *Certification data as an indication of a minimum or baseline level of qualifications.*
- *Individual transcript data on general intellectual ability and on courses taken in preparation for science or mathematics teaching–for elementary as well as secondary teachers–to provide the most complete data on teachers' formal qualifications.* The panel recognizes the cost and burdens of transcript studies but considers that such studies for samples of teachers would be valuable at the national level and to individual states.
- *Trends in guidelines for prospective teachers in terms of content or course work recommended by science and mathematics professional associations and the extent to which guidelines have been adopted.*

(b) School system factors that affect quality:

- *Hiring practices, including timing of offers, and constraints such as internal transfer rules.*
- *Teacher assignment or misassignment, by subject, including incidence of out-of-field teaching and use of temporary or emergency certification.*
- *Data describing inservice education, laboratory materials, and collegial and administrative support for teachers in place.*
- *School practices related to time use, class size, teaching load, level of autonomy, opportunities for collaboration and decision making, salary, and other monetary incentives.*

General Data Recommendations

In addition to the specific data recommendations above, certain practices should be followed to ensure the most meaningful data results and the widest use of the information.

Priority Recommendation 5. We recommend adoption of certain general guidelines for any data collection efforts relevant to teacher supply, demand, or quality:

(a) *Emphasize the repeated collection of data over time, in contrast to a one-time effort, in order to permit measurement of changes in demand, supply, and quality over time.*
(b) *Disseminate data into the public domain in a timely manner and in an easily accessible format.*
(c) *Focus on subareas of subject matter (e.g., chemistry, physics or calculus, rather than mathematics/science in general), in order to permit specific identification and targeting of areas of shortage or surplus.*

RESEARCH ISSUES IDENTIFIED BY THE PANEL

A number of important issues affecting supply, demand, and quality as they relate to science and mathematics teachers are not well understood and are beyond the scope of existing data and models. During the panel's discussions, a variety of such research topics was noted, and although they are not intended to be a comprehensive list, some of the most relevant issues are advanced for consideration by the National Science Foundation.

Resources for Research

The panel has concluded that the present research base is inadequate to support the development of behavioral models of teacher supply and demand. We therefore identify a number of issues requiring research in order to quantify the relationships needed for the development of effective behavioral models of demand, supply, and quality.

Priority Recommendation 6. The panel recommends that the National Science Foundation stimulate research on behavioral models of teacher supply and demand and increase the amount of support for such research.

Research on Demand

Policy makers frequently ask questions that could be answered by well-specified models. These include "what if" questions about the likely impacts of various education policy actions and changing labor market conditions on demand. To answer such questions, models are needed that reflect the forces that influence demand. Before such models can be developed, research is required on the behavioral factors that influence the demand for science and mathematics teachers. Although the panel's charge was to focus on supply and demand for public school teachers, changing preferences for private school enrollment, a topic about which little is known, can affect the demand for public school teachers.

Priority Recommendation 7. The panel reiterates and extends its recommendation from the interim report (National Research Council, 1987c:5-6) that research pertinent to the demand for precollege science and mathematics teachers be conducted–in order of priority–on:

(a) The behavioral determinants of student selection of science and mathematics courses at the secondary school level, including the effects of changes in graduation requirements and of student preferences for subject areas.
(b) The behavioral determinants of parental and student preferences for private and public schooling.
(c) The determinants of pupil-teacher ratios, including the adjustment lags in those ratios as enrollments change and/or the teaching force changes in demographic composition; changes in the school budget; changes in staffing patterns, typical class size, and teaching loads; increased graduation requirements; and changes in course offerings.
(d) The impact on high school dropout rates of such factors as changes in graduation requirements, labor market conditions, and the demographic composition and family circumstances of the school-age population.
(e) The relationship of changes in demand for courses to changes in pupil-teacher ratios and the resulting derived demand for full-time-equivalent teachers of mathematics and science at the secondary school level.

Research on Supply

Research on the behavioral factors that influence the supply of well-qualified science and mathematics teachers is essential to improve the understanding of teacher labor markets and to make it possible to develop dynamic models with serious behavioral content to address important policy needs.

Of prime concern is the lack of detailed knowledge of how incentives affect the supply of precollege science and mathematics teachers. Measuring the relation between supply and incentives such as salary or working conditions is important because policy makers can adjust such variables to change the supply of teachers. A related research issue concerns the supply potential of the reserve pool, which is the largest source of new entrants to teaching. Because the other source–new certificants–is decreasing in number, research to assess the supply potential of the reserve pool is of growing importance. Finally, examination of subsamples of districts experiencing supply/demand problems, including in-depth inquiries, may provide information for policy use in ameliorating the problems and can also help determine appropriate categories for disaggregation of data in publications. The recommendation below pertains to these issues.

Priority Recommendation 8. We recommend research on a variety of topics—in order of priority—that center on behavioral aspects of the supply of precollege science and mathematics teachers:

(a) *Incentives that affect individual decisions to enter teaching, to leave teaching and move to a different occupation, or to retire.*

(b) *Supply potential of the reserve pool.* Studies of the reserve pool might include the effects of incentives, such as salary increases, on attracting individuals from the reserve pool and the effects of limited mobility of teachers in the reserve pool on the supply potential of the reserve pool.

(c) *School districts experiencing supply/demand problems.* Such school districts can be identified from SASS data and studied in depth, as can the supply and demand situation in different geographic or labor market regions, e.g., inner city, rural, and high-income suburban.

Research on Quality

The pivotal role of quality in bringing teacher supply and demand into balance has proved elusive and beyond researchers' present ability to measure. We have distinguished teacher quality—referring to personal characteristics of the teacher such as education level, subject matter knowledge, skills in working with students, and degree of inservice training—from overall teaching quality. Teaching quality depends not only on teacher quality, but also on characteristics of the school and district policies that are beyond the control of the individual teacher, such as types of textbooks selected for the school system and the amount of time allocated to each subject.

In the course of panel discussion on these issues, we noted several studies related to teaching quality or teacher quality that could be pursued.

Recommendation 9. We recommend the following studies related to teaching quality and teacher quality, in order of priority:

(a) Study the effectiveness of a wide variety of practices that schools and school districts have employed to improve teaching quality in science and mathematics.

(b) Examine the inservice training practices for science and mathematics teachers that are provided by elementary and secondary schools, to identify programs that seem to be effective in enhancing teaching quality and to understand reasons why some programs appear to work while others do not.

(c) *Study teachers' transcript records, to determine the degree to which transcripts can be used as an accurate reflection of subject matter knowledge or of teacher quality.*

(d) *Study the methodological curriculum of teacher training programs to assess the degree to which these programs vary in their emphasis on pedagogical theory compared with pedagogical practice.*

Research on Student Outcomes

The ultimate usefulness of a better understanding of the supply, demand, and quality of teachers of science and mathematics lies in their effects on students' learning. It is thought that these factors are linked to outcomes, but that linkage needs to be explicit. Primary aspects of this research would attempt to relate measurable teacher characteristics, school environment variables, and home environment variables to student outcomes.

Priority Recommendation 10. The panel recommends that further research be conducted on the relationship of measurable characteristics of teachers of mathematics and science and home and school environment factors to educational outcomes of students in these fields. This research should explore variation in outcomes as well as average outcomes.

RESEARCH FACILITATION

One way to facilitate research on issues of teacher supply, demand, and quality is to ensure that the data obtained from NCES, state agencies and other studies be disseminated promptly and in a usable form to the research community. Another way to stimulate research is by providing an ongoing program of graduate student support for research. A program comparable to the National Institutes of Health training grant program in biostatistics, which was successful in attracting a large number of young researchers to the field and in changing the level of sophistication in biostatistics, could be expected to have similar effects on education statistics.

Priority Recommendation 11. We recommend that the Office of Educational Research and Improvement within the Department of Education create a program of doctoral graduate student support (training grants) in education statistics.

INFORMATION EXCHANGE AMONG DISTRICTS, STATES, AND THE NCES

The 16,000 school districts in this country operate relatively independently and balance teacher supply and demand within districts by their own actions. The staffing problems they encounter vary widely, and the actions taken by district superintendents and personnel directors to address these problems are both innovative and varied. Applicants and school systems have effective means of coping with the uncertainty of budgets and contracts and adjusting to institutional barriers (e.g., use of the substitute pool to stockpile place-bound potential teachers, use of graduate students to teach part time, and cooperative arrangements with local industry).

Much of the information about school district actions to address staffing problems will not be captured by SASS. Over the course of its study, the panel broadened its understanding of teacher supply and demand issues by direct contact with 39 public school districts across the country. They ranged from the largest metropolitan systems to the most isolated small school districts and represented a wide geographic range and a variety of labor market conditions. The case studies and the conference held by the panel with personnel directors of seven large school systems vividly demonstrated to us the diversity of practices and styles and the diversity of labor market situations that characterize the nation's school districts.

The panel believes that NCES could profit from frequent interactions with school district personnel and could play a valuable role as a broker between data producers and data users in the states. A useful mechanism for such interaction would be conferences of district and/or state personnel.

Priority Recommendation 12. The panel recommends that the National Center for Education Statistics (a) convene an annual conference of district personnel who are responsible for the decisions that affect teacher supply, demand, and quality to maintain an awareness of the current issues in teacher supply and demand; (b) hold periodic conferences of state personnel who prepare state and local supply and demand projections to facilitate improvement in these models; and (c) hold occasional conferences to promote communication between state personnel who produce data relevant to teacher supply, demand, and quality and district personnel who would find these data useful in their recruitment activities and in development of district policies concerning teachers.

IMPLEMENTING THE RECOMMENDATIONS

For immediate consideration, most of the high-priority data recommendations can be satisfied by additions to existing surveys, most notably SASS. The data elements should be added to the surveys as they are scheduled. In addition, the NCES conferences call for prompt implementation. The first annual conference should be initiated in 1990, with subsequent conferences to be planned on a continuing basis. Other recommendations for prompt implementation are the proposed program of graduate student support for studies in education statistics by the Office of Educational Research and Improvement, and the timely dissemination of data collected by NCES.

The next priorities in the time sequence for implementation are recommendations that would require new data collection instruments, such as the call for individual transcript data. In addition, a few of the data recommendations are addressed to state agencies, namely those involving certification data. The panel does not have the information to determine when it would be feasible or desirable for the states to implement these recommendations.

Finally, the research issues noted by the panel call for an expanded program of research on behavioral models of teacher supply, demand, and quality and for further stimulating this research by establishing a program for graduate student support in education statistics. The panel recognizes that this is a program of long-term research. Nonetheless, it should start immediately so that needed information and behavioral models of teacher supply and demand become available at the earliest possible date. When misconceived claims and questions about shortages are replaced with knowledge of how teacher labor markets actually function, policy makers will be able to design more sharply focused policies to ensure a strong science and mathematics teaching corps.

1
Introduction

In recent years a number of studies have expressed concern about current and prospective shortages in the nation's available supply of precollege science and mathematics teachers. Some studies claim that severe shortages currently exist; other studies find that, while current shortages are not severe, future shortages are likely; and still others find that, although there is no quantitative shortage, there is a gap between the quality of current teachers of science and mathematics and the quality needed to ensure effective instruction. Most, but not all, of the studies have focused on teachers at the secondary level, for which more information by discipline is available.

Although the panel is not charged with determining whether a shortage of precollege science and mathematics teachers either exists now or is likely in the future, the mandate to specify types of data needed to understand that issue requires the panel to examine the demographic and employment patterns affecting supply and demand in particular labor market areas. Thus we are concerned about the forces associated with changes in precollege enrollments in science and mathematics courses, including both changes in the demographic configuration of children in the relevant age ranges and changes in state or district requirements specifying the number of science and mathematics credits needed for high school graduation. We also look at the principal determinants of the total supply of teachers, including the demographics of the teacher corps.

A major concern is to understand the appropriate characteristics of teacher qualifications and teaching quality, since supply and demand for teachers come into equilibrium through adjustments in quality. Quality cannot be monitored unless the characteristics associated with it can be specified. Thus, our basic concern is to identify the types of data needed to understand quality in order to evaluate how it is changing. In the course of the effort we examine some of the available data that have led many to conclude that the quality of science and mathematics training in the United States is not satisfactory. Specifically, we examine data on student performance from studies carried out under the aegis of the International Association for the Evaluation of Educational Achievement and from the National Assessment of Educational Progress—data that have raised questions about the quality of teaching in that country.

Finally, we have considered the relationship between teacher training and preparation, teacher instructional activities in the classroom, and student outcomes. Although it is certainly true that unsatisfactory outcomes in terms of student understanding of important concepts and topics in science and mathematics can be due in part to deficiencies in the academic background or pedagogical training of science and mathematics teachers, it does not follow that poor outcomes can be attributed squarely to deficiencies in these areas.

Many factors could contribute to poor student understanding. Unsatisfactory outcomes could be due to the structure of the science or mathematics curricula; they could be due to insufficient emphasis on science and mathematics topics in the allocation of time during the school day; they could be due to the manner in which schools and classrooms are organized with respect to opportunities for interchange among teachers, the amount of time available to teachers for planning and preparation, the availability of inservice training opportunities, and so on. Poor outcomes could also be due to the fact that children receive less time and attention from parents in home environments than was true in the past, or due to changes in parents' expectations, beliefs, and behaviors related to learning science and mathematics that influence children's developmental outcomes. It is thus the panel's conviction that to understand the supply and demand for precollege science and mathematics teachers, and to understand the quality characteristics of teacher supply, we must go beyond a narrow mandate to examine the adequacy of the available data from which teacher supply and demand models could be constructed. However, the panel's mandate is not so broad that it requires us to prescribe policies whose effects might be to change either supply, demand, or quality.

THE MEANING OF *SHORTAGE*

In everyday parlance, when most people speak of a shortage of precollege science and mathematics teachers, they are likely to mean that they are dissatisfied with the quality of people teaching science and mathematics, rather than to mean that there are insufficient numbers of teachers to staff science and mathematics courses. In technical terms, it is hardly possible to have either a shortage or a surplus of particular kinds of precollege teachers, or indeed of teachers generally, since school systems typically have neither classes without teachers to teach them (excess demand/short supply) nor employed teachers without classes to teach (excess supply/short demand). Thus a quantitative shortage—fewer teachers teaching science and mathematics than there are science and mathematics classes to be taught—will not be observed except in those cases (which may be frequent but not well documented by data) in which a course or class is cancelled because a teacher cannot be found with the appropriate credentials/qualifications.

What actually takes place is an equilibrating process that is expressed in the short run by quality adjustments in the criteria for hiring next year's teachers. In the long run, salary is the equilibrating factor for supply and demand. While the quantity of people teaching science and mathematics will almost always be equal to the quantity of science and mathematics teaching offered, tendencies toward either surplus or shortage will surface as adjustments in quality. In planning for the next school year, if there are not enough applicants with science and mathematics credentials to teach science and mathematics classes, a district will either undertake aggressive recruiting or a teacher will be drafted from inside (or hired from outside) and provided with emergency certification to teach the course. If there is a potential surplus, qualified science and mathematics teachers will end up either teaching some other subject or not teaching at all. In the former case, if school systems do not recruit aggressively, they may have to dip down far into the pool of teachers less experienced or qualified in science and mathematics to fill the available positions. If the premise is true that quality is positively associated with experience and training, their average quality will tend to decline.[1] In the latter case, depending on institutional rules or practices, only the best qualified (or the most senior) science and mathematics teachers will get the available science and mathematics classes, and other teachers will have to go elsewhere or teach something else. If the available classes go to the best qualified teachers, on average the quality will tend to increase. However, if the available classes go to the most senior teacher, which is the policy in most systems, the effect on quality is difficult to assess.

[1] Some economists define commodities by listing their attributes, of which quality is one. This way of thinking about commodities, however, is a very special usage. In fact, in the situation described the district did not get first-rate mathematics and science teachers, and therefore experienced a shortage of such teachers. But it is rare that this would be revealed through questionnaire responses, since questions ask only whether the district found people who were certified in the relevant field to fill a vacancy. The dimensions in which equilibrium takes place, including quality, are relatively unobservable. The concept of shortage does not suggest a strategy for measuring shortage. Although we could refer to a "shortage of teachers of desired quality" throughout the report, for simplicity we have chosen simply to refer to a "shortage."

While these observations are straightforward and almost self-evident, they do account for the fact that some studies of the supply-demand balance in precollege science and mathematics have concluded that there is a considerable shortage of teachers, while others have concluded that no shortage exists at all. The former type of studies have defined *shortage* as the absence of sufficiently qualified teachers to staff the relevant classrooms and have judged that many classrooms are staffed by inadequately qualified teachers (as examples, see National Education Association, 1988; Weiss, 1987; Akin, 1986). The latter type of studies, asking whether schools have been unable to hire teachers to teach science and mathematics courses, have found that school systems are able to hire such teachers (as examples, see National Center for Education Statistics, 1985a; Feistritzer, 1988b). Thus the importance of the general proposition—that, although quantitative gaps between supply and demand are not generally identified, quality adjustments ensure that supply and demand are equal—is that understanding both the quantitative and the qualitative dimensions of teacher supply and demand is essential to understanding the supply and demand for teachers. To do otherwise is to miss a significant part of any potential problem.

FACTORS AFFECTING DEMAND

A data system able to track changes in the demand for precollege science and mathematics teachers must as a minimum be able to assess demographic factors, which include changes in student enrollment, in the ratio of male to female students in science and mathematics courses, and in the proportions of minority students in science and mathematics courses, as well as changes in policy variables, such as graduation requirements mandated by the state, entrance requirements of colleges and universities, and changes in acceptable pupil-teacher ratios. Demand also depends on the number of vacancies resulting from the creation of new positions and from teacher attrition. All these factors affect the demand for classes in science and mathematics.

As noted in the panel's interim report, the most accurate data used in current supply and demand models are probably the demographic data for projecting demand. For the precollege student population, projections of the total will be extremely reliable for all K-12 grades for at least five years into the future, since students starting kindergarten will already have been born about five years ago. Thus, even birth rate projections have only a small influence on demand projections, unless the projections go out further than five years. And total enrollments in grades 7-12 (the point at which specialized science and mathematics courses typically begin to be

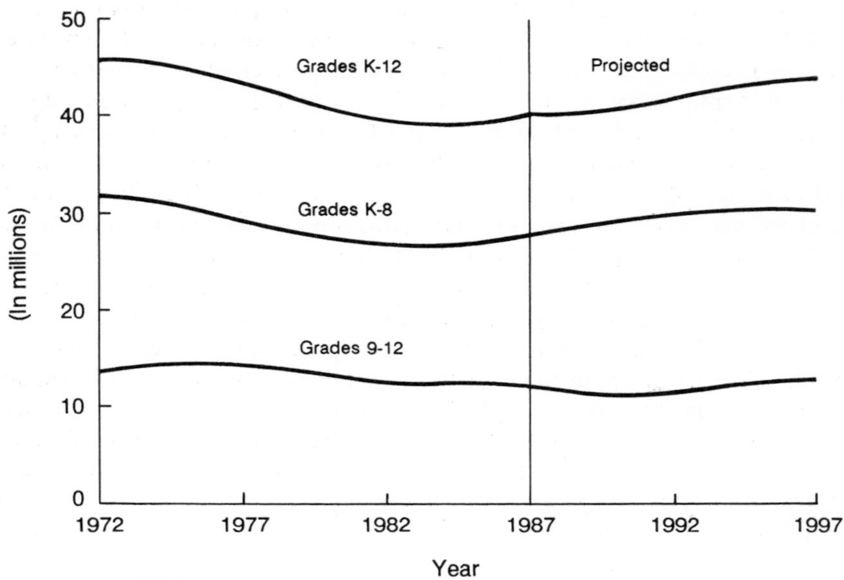

FIGURE 1.1 Enrollment in grades K-12 of public schools, with projections: Fall 1972 to 1997. Source: National Center for Education Statistics (1988g:14).

offered) are known at least 12 years in advance since the children have already been born.

The demographic base for projecting demand is not quite so solid as the above paragraph suggests, even at a national level. Both in-migration and out-migration occur among school-age children in the United States as a whole. But at the more relevant regional, state, or local level, there is obviously some migration of school-age children that must be taken into account statistically. Thus, even very good national models need to be substantially augmented with accurate subnational migration data to produce useful demand projections at the relevant school district level.

School enrollment itself is projected to rise somewhat during the next half-decade. For example, from 1978 to 1987, enrollment in public secondary schools (grades 9-12) fell from about 14 million students to about 12 million; but from 1988 to 1995 secondary enrollment is projected to increase to about 13 million. Over the same period, elementary school enrollment is projected to increase from a little over 28 million students to about 31 million (see Figure 1.1).

A number of forces currently under way suggest the need to track more refined characteristics of the demographics and to add to them data relating to mandated state requirements.For example, a potential exists for greater demand for science and mathematics training for female children in the K-12 age range, simply because of persistent changes in attitudes toward appropriate sex roles for men and women and the associated changes in the career aspirations of young women. It is already evident that more young women are planning to enter science and engineering fields than was true 20 years ago. In fact, the number of undergraduate women majoring in science and engineering has risen dramatically since the mid-1970s. For example, the physical sciences showed an increase from 30,900 women in 1976 to 38,100 in 1984, then decreased to 36,500 in 1986 (National Center for Education Statistics, 1988b:167). Similarly, in 1976, 28,800 women majored in engineering. This number increased to 74,800 by 1984 and declined slightly to 71,200 in 1986. Though trends have attenuated for the present, it is important to monitor the enrollment of women as science and mathematics majors at the postsecondary level. Currently, 28 percent of all physical sciences majors are female; because of the potential for further increase, female enrollment in science and mathematics should be monitored.

Similarly, it seems likely that the movement toward equal opportunity will generate an increased demand for science and mathematics training on the part of minority youngsters. The evidence here, some of which is shown in Tables 1.1 and 1.2, is not easy to interpret. For students who were high school seniors in 1980, the 1980 data indicate that blacks actually took more semesters of mathematics than whites, and only Chicanos and Native Americans (especially the latter) are markedly lower than average. For science, black seniors were well below whites in number of semester hours in 1972, but in 1980 black seniors had nearly caught up with whites, while all the other minority groups except Asian-American and Puerto Rican students were below whites. These data are for seniors, and high school dropout rates are much higher for minority students than for whites. Moreover, the data noted above do not standardize for the level of science and mathematics courses. Minorities other than Asian-Americans are historically more likely to be found in remedial mathematics than in the more challenging mathematics courses and in general science courses than in physics and chemistry (Office of Technology Assessment, 1988:45).

A recent report by the Educational Testing Service (ETS) drawn from a research paper on course-taking patterns in the 1980s by Goertz (1989), compares students' course-taking patterns in 1982 and 1987, using data from the High School and Beyond study by the National Center for Education Statistics (NCES) (1982 graduates) and the High School Transcript Study by Westat, Inc. (1987 graduates). This report (Educational Testing Service, 1989:20) finds significant gains in course-taking by black and Hispanic students between 1982 and 1987. Some of the gains were impressive, others modest. For example, 29 percent of black graduates in 1982 had

TABLE 1.1 Semesters of High School Mathematics (Grades 10-12) Reported by Seniors, by Racial/Ethnic Group, 1972 and 1980

	1972 National Longitudinal Study			1980 High School and Beyond Study			
	Mean	Standard Deviation	Sample N	Mean	Standard Deviation	Sample N	Change 1972-1980
TOTAL	3.93	1.8	11,771 [a]	4.06	1.9	27,928	0.14*
White	3.97	1.8	9,228	4.04	1.9	19,695	0.07
Black	3.86	1.6	1,402	4.28	1.8	3,709	0.42*
Native American	2.67	1.7	117	3.52	1.9	215	0.85*
Chicano	3.30	1.7	341	3.73	1.8	1,873	0.43*
Puerto Rican	4.09	2.0	62	4.26	2.1	303	0.18
Other Hispanic	4.12	1.7	77	4.07	1.9	959	-0.05
Asian-American	4.28	1.8	144	4.91	1.8	364	0.63*

* p less than or equal to 0.05

[a] Item response number is smaller than number of survey respondents.

Source: Rock et al. (1985: Table 6-48).

TABLE 1.2 Semesters of High School Science (Grades 10-12) Reported by Seniors, by Racial/Ethnic Group, 1972 and 1980

	1972 National Longitudinal Study			1980 High School and Beyond Study			
	Mean	Standard Deviation	Sample N	Mean	Standard Deviation	Sample N	Change 1972-1980
TOTAL	3.71	1.8	12,002 [a]	3.46	1.9	27,482	-0.25*
White	3.77	1.8	9,397	3.48	2.0	19,465	-0.29*
Black	3.52	1.7	1,449	3.45	1.9	3,604	-0.02
Native American	2.75	1.5	120	3.02	1.7	212	0.27
Chicano	2.96	1.6	350	3.05	1.6	1,828	0.09
Puerto Rican	3.67	1.7	66	3.57	2.0	290	-0.10
Other Hispanic	3.80	1.9	81	3.31	1.8	945	-0.48
Asian-American	3.82	1.8	138	4.12	1.9	357	0.30

* p less than or equal to 0.05

[a] Item response number is smaller than number of survey respondents.

Source: Rock et al. (1985: Table 6-50).

taken geometry; by 1987, 44 percent had taken geometry, compared with 64 percent of whites. In calculus, gains were slight for blacks, from 1 percent in 1982 to 2 percent in 1987, compared with 6 percent for whites. The percentage of Hispanic graduates who had taken algebra I rose from 55 to 77 percent between 1982 and 1987, and by 1987 they were nearly even with whites (at 78 percent). Minority gains in science course-taking were similarly notable. However, blacks and Hispanics still lag behind whites and Asians in their enrollments in the higher-level mathematics and science courses.

In addition, school districts have changed their graduation requirements to include more science and mathematics training or credits required for graduation from high school. In 1985, NCES surveyed a sample of 565 districts and asked for math and science requirements for high school graduation in 1982, 1985, and the expected requirements in 1988. Between 1981-82 and 1984-85, for example, nationally the average number of years of course work required for graduation from public high schools increased from 1.6 to 1.9 for mathematics, and from 1.5 to 1.8 for science (see Table 1.3). The National Commission on Excellence in Education (1983) recommended 3.0 years for both science and mathematics.

In response to changes in the graduation requirements of districts, states, and even in the entrance requirements of colleges and universities, increased enrollments in high school science and mathematics courses have been documented (Educational Testing Service, 1989). The years between 1982 and 1987 have seen strong gains in science and mathematics course-taking, except in physics and calculus, for which gains were modest or nonexistent. To the extent that new state course requirements exceed those already in place in the districts, the result can be a stronger demand for science and mathematics training, given the same student population. However, when local school district requirements already exceed new state requirements, which they often do, new demand for teachers may not result. Therefore it is important to monitor changes in course requirements at both the state and district level to assess the effects on the demand for teachers.

In addition to changes in course requirements, a number of other policy-related factors influence the demand for new science and mathematics teachers. Changes in pupil-teacher ratio can result in changes in demand. And a number of policy-related factors at the school, district, or state level can influence the ratio—changes in budgets, class size policies, or course requirements, for example. These changes should be monitored in any data system that tracks changes in demand for science and mathematics teachers.

Another major component of demand models is the pattern of attrition for science and mathematics teachers—due both to retirement and

TABLE 1.3 Average Years of Course Work Required for High School Graduation by Public School Districts With High Schools, 1981-82, 1984-85, and 1987-88

	Subject Area				
School Year of Graduation	Mathematics	Science	English	Foreign Languages	Social Studies
1981-82	1.6	1.5	3.6	0.0 [a]	2.6
1984-85	1.9	1.8	3.8	0.1	2.8
1987-88 [b]	2.3	2.0	3.9	0.2	2.9
Recommendations of National Commission on Excellence in Education [c]	3.0	3.0	4.0	2.0 [d]	3.0

[a] Less than 0.05 years.

[b] Expectations as of fall 1985 about requirements for seniors graduating in 1988.

[c] Another half year of course work was recommended in computer science. Almost no school districts had requirements in this area in 1981-82. That situation changed by 1984-85, when the average for all school districts was 0.1 years of course work required for graduation in computer science; the expected average for 1987-88 is 0.2 years.

[d] The recommendations of the Commission on Excellence in Education about foreign languages applied only to the college-bound, not to all students. The figures for actual requirements represent requirements for all graduates.

Source: Center for Education Statistics (1987a:84).

especially to leaving earlier in one's teaching career. It is essential for an effective data system to be able to monitor attrition rates by subject as well.

Finally, research is called for to identify other behavioral factors that influence the demand for teachers: for example, patterns of dropping out of high school, parents' choice of private over public schools, and the timing of that choice.

FACTORS AFFECTING SUPPLY

Teacher supply can be examined in terms of retention rates for the present stock of teachers, the flow of newly certified teachers from colleges and universities, and the flow of returning teachers who have been absent from the labor market, laid off during the past decade due to declining enrollments, or have come from other occupations or alternative certification routes. As with demand, these factors include both demographic characteristics (the age distribution of current teachers) and policy variables. In our interim report, we noted that most of the existing supply models focus on the flow of new graduates of education degree programs, despite the fact that most of the new hires during recent years have come from other sources.

The existing data, most commonly from the states and from periodic surveys at a national level, should be examined in greater detail to estimate future declines in the supply of available teachers, both for precollege science and mathematics and for precollege teachers generally. To what extent will there be a substantial decline in the overall teacher retention rate, arising from the fact that large numbers of teachers will be entering the age and experience combination at which teachers have often retired in the past? One of the best-established relationships in the teacher supply literature is the U-shaped relationship between age/experience and teacher retention: in the early years, attrition rates are high—either because many entering teachers find that the occupation is not what they had thought, have adverse experiences that result in withdrawal from the teacher corps, or find more attractive employment opportunities. At the other end of the spectrum, where the older and more experienced teachers are located, attrition rates rise as retirement approaches. Table 1.4 illustrates this pattern for the state of New York. Both early and late attrition estimates will be important factors affecting the supply of science and mathematics teachers.

The current composition of the teacher corps is concentrated in an age/experience cohort in which there will be many retirements starting in the late 1990s. For example, the 1985-86 Survey of Science and Mathematics Education conducted for the National Science Foundation (Weiss, 1987) found some indication that the science and mathematics teaching

TABLE 1.4 Public School Classroom Teacher Attrition and New Hires by Age Group: New York State, Fall 1985 to Fall 1986

Fall 1985 Total Classroom Teachers (N = 175,256)		Attrition Between Fall '85 & Fall '86 (N = 15,598)		New Hires in Fall 1986: Beginning (1st Year) Teachers (N = 4,001)		New Hires in Fall 1986: Experienced Teachers (N = 12,462)		Fall 1986 Total Classroom Teachers (N = 176,121)	
Age	Vertical Percent	Age	Vertical Percent	Age	Vertical Percent	Age	Vertical Percent	Age	Vertical Percent
<35	22.0%	<35	29.0%	<35	78.4%	<35	1.5%	<35	21.1%
35-39	23.4	35-39	16.6	35-39	11.3	35-39	2.3	35-39	21.9
40-44	19.5	40-44	11.3	40-44	5.5	40-44	37.3	40-44	21.3
45-49	13.0	45-49	7.9	45-49	2.4	45-49	19.4	45-49	13.8
50-54	11.3	50-54	10.9	50-54	1.3	50-54	12.4	50-54	11.2
>55	10.8	>55	24.3	>55	1.1	>55	27.1	>55	10.7

Source: New York State Education Department (1989). Files of the Information Center on Education.

force is aging but did not predict an unusually large wave of retirees in the near future. Monitoring relevant statistics related to factors associated with choosing to stay or choosing to retire is crucially important for understanding future teacher supply, as is monitoring the effects of incentive programs designed to encourage continuation and discourage retirement (or vice versa).

One sign of an impending shortage of new teachers has been a decline in the number of education degrees awarded. For example, the number of bachelor's degrees in education fell from 108,000 in 1980-81 to 87,000 in 1985-86 (National Center for Education Statistics, 1986:134; 1988b:196). The number of master's degrees in education also declined, from 99,000 in 1980-81 to 76,000 in 1985-86. Among those enrolled as teacher candidates in secondary education programs, the proportion majoring in mathematics education held steady at about 25 percent between 1985 and 1988. However, the proportion of students majoring in science education has declined from 21 percent of all enrolled in 1986 to 16 percent in 1988 (AACTE, 1989). The shortage issue is complicated, since new teacher supply can be fairly quickly adjusted as opportunities are perceived to arise.

Conventional teacher training institutions are not the only source of new supply. In recent years new supply has come mainly from a broader source of teachers that includes (1) graduates of other institutions who enter the teacher supply with temporary credentials and later certification; (2) the so-called reserve pool: past graduates of teacher training or other institutions who did not enter teaching when they graduated but could be attracted to teaching careers with the right incentives; and (3) former teachers who return to teaching from another occupation or activity. In short, monitoring the basic demographics of teacher age/experience, as well as the potential supply of new graduates and returnees from other occupations, will be crucially important to understanding the probable evolution of teacher supply over the next decade.

QUALITY ISSUES IN SUPPLY AND DEMAND

Much of the impetus for concern over the supply, demand, and quality of precollege science and mathematics teachers arises from the continuing evidence that U.S. students do not appear to know as much science and mathematics as their age peers in other countries. The most widely cited such data come from the International Educational Assessment program (IEA) and from the National Assessment of Educational Progress (NAEP).

The IEA administered science tests to fifth-grade and ninth-grade students and to twelfth-grade students who were studying biology, chemistry, or physics in the terminal grade in school in 17 countries in 1983 (1986 for the United States). The results of these tests tend to show U.S. students' science performance declining from a middle position in fifth grade to quite low by twelfth grade (IEA, 1988). In science, U.S. 10-year-olds were eighth among 15 countries ranked, U.S. 14-year-olds were fourteenth among 17 countries ranked, and of 13 countries ranked for twelfth-grade students who were taking science courses, U.S. biology students were thirteenth, chemistry students eleventh, and physics students ninth (Table 1.5). In general, although U.S. students did relatively poorly overall, they did worse at the higher grades and better at the lower grades.

This may be explained in part by cross-national differences in science curricula. The science curricula in the other countries participating in this study generally require more years of science than are required in the United States. The U.S. results for grade 12 generally correspond to student achievement near the end of their second year of the subject; students in the other countries generally would have completed three years of the science by grade 12 (Jacobson and Doran, 1988).

U.S. twelfth-grade college-preparatory mathematics students fared poorly against their peers in both developed and less-developed countries of the world in performance on mathematics achievement tests (McKnight et al., 1987). For example, for high school seniors taking mathematics, U.S. students' scores ranked in the lowest quarter of the countries in three categories (number systems, algebra, and geometry) and were below the median in the other three (sets and relations, elementary functions/calculus, and probability/statistics). Eighth-grade students in the United States ranked somewhat higher, scoring at the median in arithmetic, algebra, and statistics; at the 25th percentile in geometry; and below it in measurement. The mathematics data are shown in Tables 1.6 and 1.7.

In the IEA mathematics study, the method used by the United States, England, and Wales to obtain a sufficiently large number of cooperating school districts, namely requesting participation of twice as many school districts as were needed with the expectation of a 50 percent cooperation rate, might be expected to produce a bias in achievement scores. However, no evidence of bias has been found (Garden, 1987:133).

Neither of the international comparisons is without its problems and ambiguities. For example, it is not clear whether the student populations tested in the IEA science study are fully comparable across countries. Furthermore, it is sometimes argued that the tests themselves are biased, since the U.S. curriculum in science and mathematics may be different from the typical curriculum used elsewhere, and the tests may be heavily weighted with items that are not covered in U.S. curricula. Data from the IEA study on test validity and test relevance, however, do not support that proposition (IEA, 1988:88-95).

A similar picture is presented by the NAEP data on achievement scores, which indicate that large fractions of U.S. students do not appear

TABLE 1.5 Rank Order of Countries for Science Achievement at Three Levels of Schooling

	10-Year -Olds, Grade 4/5	14-Year -Olds, Grade 8/9	Grade 12/13 Science Students [a]			Non-Science Students
			Biology	Chemistry	Physics	
Australia	9	10	9	6	8	4
Canada (English speaking)	6	4	11	12	11	8
England	12	11	2	2	2	2
Finland	3	5	7	13	12	-
Hong Kong	13	16	5	1	1	-
Hungary	5	1	3	5	3	1
Italy	7	11	12	10	13	7
Japan	1	2	10	4	4	3
Korea	1	7	-	-	-	-
Netherlands	-	3	-	-	-	-
Norway	10	9	6	8	6	5
Philippines	15	17	-	-	-	-
Poland	11	7	4	7	7	-
Singapore	13	14	1	3	5	6
Sweden	4	6	8	9	10	-
Thailand	-	14	-	-	-	-
U.S.A.	8	14	13	11	9	-
Total Number of Countries	15	17	13	13	13	8

[a] Students taking biology, chemistry, or physics in the terminal grade in school.

Source: International Association for the Evaluation of Educational Achievement (1988:3).

TABLE 1.6 Mathematics Achievement Comparisons: Twelfth Grade United States and International, 1981-82 (Percentage of Items Correct)

	United States			International (15 Countries)		
Topic	Pre-calculus Classes	Calculus Classes	Total	25th Percentile	Median	75th Percentile
Sets & relations	54	64	56	51	61	72
Number systems	38	48	40	40	47	59
Algebra	40	57	43	47	57	66
Geometry	30	38	31	33	42	49
Elementary functions/ calculus	25	49	29	28	46	55
Probability/ statistics	39	48	40	38	46	64

Source: McKnight et al. (1987:23)

to meet minimal standards of literacy in science and mathematics. The NAEP *Science Report Card* of September 1988 indicated that, despite gains over the past four years, particularly among minorities, a majority of high school students "are poorly equipped for informed citizenship and productive performance in the workplace" (National Assessment of Educational Progress, 1988b:5).

A problem with both NAEP and the IEA tests is the limited extent to which they assess higher-order skills. Although some test materials administered by NAEP and IEA involve hands-on exercises, much more research and development activity is needed to construct free-response materials and techniques that measure skills not measured with multiple choice tests. Current improvements in mathematics and science curricula are focused on learning of "conceptual knowledge, process skills, and the higher-order thinking that scientists, mathematicians, and educators

TABLE 1.7 Mathematics Achievement Comparisons: Eighth Grade, United States and International, 1981-82 (Percentage of Items Correct)

	United States	International (20 Countries)		
Topic	(Percentage Correct)	25th Percentile	Median	75th Percentile
Arithmetic	51	45	51	57
Algebra	43	39	43	50
Geometry	38	38	43	45
Statistics	57	52	57	60
Measurement	42	47	51	58

Source: McKnight et al. (1987:21).

consider most important" (Murnane and Raizen, 1988:63). It is not clear what the relative standing of U.S. students would be on a test that assessed higher-order thinking skills more fully.

Despite all the caveats that have been and can be made with regard to these comparisons, the evidence is that U.S. high school students cannot be judged to perform well in science or mathematics by any reasonable standard, or at least not as well as society seems to expect.

Evidence from IEA that young people who concentrate heavily in science and mathematics do not perform especially well implies even worse outcomes for the great majority of American youth who take very little science and mathematics in high school. From the perspective of employers, for example, what matters at least as much as the quality of instruction for high school students who are potential scientists and engineers is the quality of technical or quantitative training for the great majority of high school students who will not go on to these types of careers but will enter the work force after graduation. Concern over the ability of young people to function effectively in today's technical environment, given the inadequacy and often the total absence of science and mathematics training with any degree of rigor, looms as a major societal concern and is the subject of numerous recent reports.

Both low test scores and the generally low level of scientific literacy underpin the concern with the quality of science and mathematics training, and with the prospective shortage of qualified science and mathematics teachers. Poor outcomes have thus spurred a deep concern with the quality of teaching and the qualifications of teachers of science and mathematics. Since it is through adjustments in quality that the supply and demand for precollege science and mathematics teachers reach equilibrium in the short run—e.g., the next school year—an examination of possible statistics to measure quality is of central concern to the panel.

At least two different sets of factors are relevant to an assessment of teaching quality. One set relates to the teaching environment and includes school, district, and state policies and practices that enhance or impede one's ability to secure the right teaching assignment and to teach effectively. Such factors include time spent on science and mathematics, teaching burden, textbook use, district decisions about recruiting and hiring teachers, and inservice education policies.

Another set of factors relates to the background and qualifications of the individual teacher. These include type of certification, relevant courses taken in the past and currently, and measures of cognitive ability. The need for better data on these kinds of factors, both for monitoring supply and demand and for modeling purposes, is discussed in Chapter 5.

It should be kept in mind that even if all the comparison data were valid and indicated that U.S. students have low absolute and relative achievement in science and mathematics, it would not necessarily follow that the problem lies solely or even mainly with the training of U.S. teachers of precollege science and mathematics. Educational outcomes are a complex function of student and family inputs, teaching inputs, educational curricula, school and community environment factors, and student behaviors, including student such as doing homework, attitudes toward science and mathematics, and scientific habits such as objectivity, skepticism, and replication of results (Murnane and Raizen, 1988). Poor outcomes can clearly be due in part to the inadequate training of teachers, but they can also be due to factors that have little or nothing to do with the training and ability of the teacher corps.

For example, there has been a continuing dispute among mathematics teachers about curricular issues, which are seen by some as having a strong influence on the level of performance of U.S. students in standardized tests of mathematics skills. It is alleged that mathematics skills in U.S. schools are typically taught in a layered or "spiral" curriculum, whereby students are taught a number of concepts in grade t, and are then taught slightly augmented but basically similar concepts in grades t+1, t+2, ... It is argued that students are thus introduced to relatively little new material each year through grade 8, that most of what is done constitutes review of materials previously taught, and that as a result students become bored with the constant repetition and never really master many of the key ideas involved in the development of mathematical skills. It is also judged by people who hold this view that part of the problem is that mathematics textbook producers try to widen the appeal of their product to as many school systems as possible; they end up including small segments on a variety of topics and intensive treatment of few, if any, of these topics. Since the basic text is the primary resource used by most precollege mathematics teachers (Weiss, 1987:31, 39), and since the text usually favors breadth and facts over depth (Office of Technology Assessment, 1988:30-34), the result is that a significant fraction of students master few if any of the topics.

A different line of argument, which could in principle be resolved more easily and might make a substantial difference to outcomes, is that U.S. students have inadequate skills in science and mathematics simply because teachers, especially elementary school teachers, do not spend much classroom time on science and mathematics topics. Research indicates a great deal more time is devoted to reading than to mathematics (Cawelti and Adkisson, 1985; Weiss, 1987). Observations of actual classroom time spent on mathematics also have found very large differences between students in Minneapolis, Minnesota, and those in Taipei, Taiwan, or Sendai, Japan (Stevenson et al., 1986): U.S. students spend far less time on mathematics than do Asian students.

To the extent that the performance of U.S. students on science and mathematics tests and the level of their skill in these areas is simply due to the emphasis on language arts found in U.S. classrooms and/or to the smaller amount of time spent either in school or in school-related activities at home, both the interpretation of the problem and the solution are relatively simple—provided school systems can be encouraged or induced to change the structure of their curricula. But if that is the basic problem, then the issue again is not one of inadequacy of preparation or academic training on the part of teachers of science and mathematics in the United States, but simply one of relative emphasis within the curriculum. In that case, the question should be raised as to why fewer hours are spent on science and mathematics in American classrooms.

Of course, it is possible that one reason U.S. students spend less time on science and mathematics is that many U.S. elementary school teachers are much less comfortable in teaching science and mathematics than in teaching language arts, and that part of the reason for the curricular emphasis is a preference on the part of teachers and/or administrators derived in turn from their own training. It appears unlikely that specialist teachers of mathematics in the early grades would be motivated to shorten class time spent on mathematics, and use of such teachers is more common in Japan, China, and Taiwan than in the United States in the early grades.

There also appear to be differences in the nature of the pedagogical training given to U.S. and Japanese mathematics teachers. The IEA mathematics study (McKnight et al., 1987) reported that among mathematics teachers at the eighth- and twelfth-grade levels, the U.S. teachers had taken more mathematics courses and fewer mathematics pedagogy courses than their Japanese counterparts (p. 64). American teachers also have much less nonteaching time scheduled during the day, compared with their Asian counterparts (Stevenson, 1987:32). And the degree of teacher autonomy is different: U.S. teachers are often on their own after the first year, while in Asian classrooms younger teachers are typically under the tutelage of a senior teacher for a number of years (Lee et al., 1987; Stevenson et al., 1988; Stigler et al., 1987; Stevenson and Bartsch, in press).

Home environment is another factor that affects student outcome. There is considerable evidence that learning and training for young children take place in the home as well as in the school, and that the relative importance of training in the home is much greater when children are young. The home environments in which children are being raised in the United States are considerably different now from the way they were several decades ago. The proportion of children raised in single-parent households is much larger now than in the past, and the proportion of mothers who work full- or part-time is much higher now than in past decades. These realities can create problems for children, especially for minority children, many of whom are raised in single-parent households for a substantial portion of their developmental years (Hill et al., 1987). Although we cannot be certain that the amount of time and attention parents pay to young children's development is necessarily less because there are fewer "parent hours" available in the aggregate, it is certainly plausible to suppose that fewer total parent hours will result in fewer developmental hours spent by parents on children. There is some evidence that working mothers largely trade off leisure time and sleep for work hours, not for time spent with their children (Hill and Stafford, 1985). In any event, demographic characteristics have a potentially serious influence on the process of skill development in young children, and part of understanding educational outcomes is surely to understand how these home environment factors relate to these outcomes.

In addition to the demographic differences in home environments, there also appear to be substantial differences in the practices, beliefs, and expectations of parents in American households compared with those in other countries. Again, the best-documented evidence comes from a comparison of American and Asian households. As a generalization, Asian mothers are less satisfied with the school performance of their children than American mothers (despite the fact that their children are generally doing better), they are more likely to attribute success in school to hard work rather than to native ability, and they are less likely to be satisfied with the way the schools are performing than their American counterparts (Lee et al., 1987).

The implication of the issues just discussed is not that the solution to poor performance on standardized educational outcome tests, and presumptively in the level of skill development in science and mathematics for American students, are to be found in factors other than either the quantity of precollege science and mathematics teachers or the quality of their training characteristics or classroom methods. Rather, it is that poor student outcomes are not uniquely correlated with, nor necessarily caused by, inadequate quantity or quality, but could easily be due to factors that are largely unrelated to teacher or teaching quality. It would thus be a mistake, in the panel's view, to jump to the conclusion that poor science and mathematics outcomes on the part of students necessarily reflect inadequacies in the background, training, or ability of their teachers and to seek the remedy for the problem only by enhancing either the numbers or the quality of precollege science and mathematics teachers. That could turn out to be the case, but many other factors, such as the structure of the curriculum, the practices of both K-12 school systems and teacher training institutions, the amount of time spent on science and mathematics topics in schools, and the influence of home environments on development outcomes, all need to be understood before we can expect either to understand the problem or to devise appropriate remedies.

THE PANEL'S WORK AND ORGANIZATION OF THE REPORT

During the course of its work, the panel broadened its understanding of the flow of teachers through school systems by direct contact with 39 public school districts across the country. These school systems ranged from the largest metropolitan systems to the most isolated small school districts and represented a wide geographic range and a variety of labor market conditions.

Six of the 39 districts were the subject of in-depth case studies, conducted in 1987 and 1988, of supply and demand issues regarding science and mathematics teachers. Two of the districts were in California and near one another geographically: one was a large urban system whose ability to attract talented science and mathematics teachers was affected by a history of budgetary constraints and teacher-organization or school-district provisions, while the other, a small, wealthy district, was able to exercise greater autonomy in attracting and keeping talented teachers. Two other districts—one in California and one in Utah—both with large enrollments, were selected for their growing populations and rising enrollments. Hiring in one of them was severely limited by fiscal constraint as well as strong religious and community standards; the other was growing in both enrollment and economic base. Finally, two contiguous school districts in Maryland that were expected to hire from the same labor market were visited—a large urban district coping with school closures, leadership changes, and traumatic layoffs as the student population has moved to the suburbs and a medium-sized, stable, semirural school district nearby.

The in-depth case studies furnished invaluable context without which statistics portraying supply and demand would be seriously incomplete. Such context showed the role of the individual personnel administrator and his or her ability to maneuver or use informal networks to attract science and mathematics teachers. It showed the effects of competing labor markets, teacher-organization provisions, budgetary constraints, and other external factors.

The six in-depth case studies were supplemented by 27 additional mini case studies, conducted by telephone interviews and follow-up questionnaires, in order to test the representativeness of the findings. The mini case studies were conducted over the period June through December 1988.

Finally, a conference of the chief personnel administrators of seven large metropolitan school districts, representing over 5 percent of the nation's total public school enrollment, was convened in May 1988.[2] Issues of supply, demand, and quality of science and mathematics teachers were discussed, and the districts' statistical information systems were examined for data relevant to supply and demand models. Appendix A provides more information about each of these activities. Discussions in the chapters that follow frequently draw on the experiences of the school district personnel administrators who participated in these studies.

In the chapters that follow, we further examine the characteristics of demand for precollege science and mathematics teachers (Chapter 2) and issues relating to supply (Chapters 3 and 4). Chapter 3 reviews projection and behavioral models and the essential behavioral components of effective supply models. It examines individual incentives to teach and school district actions that influence supply decisions and mesh supply with demand. In Chapter 4, data needed to monitor the supply pool along its various stages in the teaching career are discussed.

We then turn to the role of quality adjustments in bringing supply and demand to equilibrium. In Chapter 5 we look at the question of measuring teacher characteristics and teaching quality. Chapter 6 contains the panel's conclusions and recommendations: some of the recommendations deal with specific data needed to better understand demand, supply, or quality

[2]The number of districts in the three case study activities add up to 40, with one of the large school districts participating in two of the projects.

factors, and other recommendations deal with the types of research needed to better understand the linkages among demand, supply, teacher quality, and student outcomes and ways to facilitate this research.

2
Determining Teacher Demand

In this chapter we are concerned with the demand for new teachers, specifically teachers new to a particular job. Projections of the demand for new teachers require the projection of a minimum of three data elements: student enrollment, pupil-teacher ratios, and teacher attrition rates. Demand projections for segments of the teacher population, such as secondary school teachers of mathematics and science, require the projection of these data elements for the specific segments. For example, projecting the demand for mathematics teachers requires, at a minimum, projections of enrollment in mathematics classes, the expected size of mathematics classes, and the attrition rate of mathematics teachers. The necessary data for teacher demand projections vary in availability and reliability. Pupil-teacher ratios vary with staffing patterns, class sizes, teaching loads, course requirements, and course-taking patterns in science and mathematics. Meaningful projections of the consequences of course-taking patterns or teacher attrition are typically less available than are projections of future student enrollment. Statewide enrollment projections are more reliable than those for local school districts. In general, the smaller the subset projected, the lower the reliability.

STUDENT ENROLLMENT

The two main topics discussed in this section describe methods of projecting student enrollment, based on either public-school enrollment data or population data. The main features and limitations of both sources of data are discussed. We then turn to the other key components of estimating demand for teachers—pupil-teacher ratios and attrition.

Enrollment Projections Based on Student Enrollment Data

Projections of student enrollment, one of the three elements necessary for projecting the demand for new teachers, are the easiest. K-12 enrollment projections are widely available. Most states and many local districts produce enrollment projections, particularly for the public schools, although some states produce them for both public and private schools. These projections typically follow a standard "cohort survival" methodology, which uses observed enrollment ratios between grades to move ("survive") classes ("cohorts") forward to the next level. If the state or school system has low or constant levels of migration, then reliable projections by grade are produced.

The National Center for Education Statistics (NCES) produces one-year projections by state and nationwide projections for 10 years into the future. Until 1988, NCES projected enrollments were for the public schools only; projections for private schools began in 1989. NCES employs a mixed model in which participation rates for kindergarten, grade 1, and special and ungraded classes are calculated by applying recent public school enrollment data, collected in its annual survey of the states, to age-specific population estimates produced by the Census Bureau. The resultant rates are then applied to projected populations of the appropriate ages (e.g., 5-year-olds for kindergarten, 6-year-olds for first graders) to arrive at levels of future enrollment for those grades and classes. Retention or grade progression rates from the NCES annual survey are used to calculate grades 2 through 12. The NCES method of projecting enrollments is described more fully in Part III of the panel's interim report (National Research Council, 1987c), which discusses the components of the NCES model and those of six states. The sources of data used by NCES are also described in the Guide to Sources portion of the *Digest of Education Statistics* (National Center for Education Statistics, 1988b:358-380).

Enrollment-based changes in the demand for public school teachers below the national level are usually better obtained directly from actual enrollment projections for the state or locality in question than from projections of the school-age population. The reason for preferring enrollment projections at the subnational level is the difficulty of projecting internal and foreign migration for subnational populations. Although birth and death data for population projections are quite accurate, migration estimates are less certain, especially as they must be allocated by age. Both the uncertainty and the effect of migration are greater for subnational aggregations than for the entire country, yet no data are collected for interstate or intrastate movement.

Enrollment projections, however, are typically based on annual censuses of the school population that are taken for administrative purposes, including the allocation of state support. The bases for enrollment projections are therefore firm, and the grade progression ratios can be updated annually as circumstances change. The changing circumstances include net migration, which is picked up by the grade progression ratios.

Geographic Differences in Projected K-12 Enrollment

There are great differences among regions of the United States, and the localities within those regions, in prospective public school enrollment change over the next 10 to 20 years. The nationally projected growth, and expected eventual decline, in the 5- to 13-year-old population and the current nationwide decline and subsequent slow growth in the 14- to 17-year-old population are far from evenly distributed among states or within states.

A well-known compendium of individual state enrollment projections that shows the dramatically different demographic expectations among states is produced by the Western Interstate Commission for Higher Education (WICHE) in cooperation with Teachers Insurance and Annuity Association and the College Board. WICHE produces projections of numbers of high school graduates using K-12 enrollment data provided by the states and a cohort survival methodology. The authors do not attempt to integrate the individual state projections into a valid national projection by making explicit assumptions about migratory movements among the states, although they do sum the projections into regional and a national totals. However, recent migratory movements are embedded in the observed progression ratios for each state, which are used to move the enrollments forward into the future. The WICHE projections illustrate the potential for sharply divergent demographic futures in the different regions of the nation and within states between 1986-91 and 2003-4.

In WICHE's 1988 set of projections (WICHE, 1988) the number of high school graduates nationally has formed a "roller coaster" pattern since the late 1970s, a pattern that will continue through the next decade. The general roller coaster pattern reflects past birth patterns in the United States, but it differs from region to region and from state to state. The regional differences, the document's foreword explains, are due to "the mobility of the population, varying economic conditions, and growth in minority populations." The number of high school graduates is projected to decline for the North Central and Northeast states, while the West, South, and South Central regions are projected to have little decline in the mid-1990s and then will experience substantial growth (Tables 2.1 and 2.2).

TABLE 2.1 Projections of Numbers of High School Graduates, by Region, 1986-2004

	West		South/ South Central		North Central		Northeast		United States	
	Total	Index (1985 -86 = 1.00)	Total	Index (1985 -86 = 1.00)	Total	Index (1985 -86 = 1.00)	Total	Index (1985 -86 = 1.00)	Total	Index (1985 -86 = 1.00)
1985-86 (actual)	484.087	1.00	787,625	1.00	725,154	1.00	653,576	1.00	2,650,442	1.00
1986-87	491,388	1.02	811,738	1.03	736,070	1.02	655,907	1.00	2,695,102	1.02
1987-88	521,196	1.08	828,325	1.05	754,949	1.04	663,718	1.02	2,768,189	1.04
1988-89	511,390	1.06	844,410	1.07	743,365	1.03	633,419	0.97	2,732,584	1.03
1989-90	489,118	1.01	816,646	1.04	699,696	0.96	588,978	0.90	2,594,438	0.98

1990-91	476,320	0.98	789,721	1.00	658,301	0.91	549,690	0.84	2,474,032	0.93
1991-92	482,357	1.00	776,433	0.99	646,083	0.89	536,181	0.82	2,441,054	0.92
1992-93	491,783	1.02	775,419	0.98	649,665	0.90	533,038	0.82	2,449,905	0.92
1993-94	507,053	1.05	773,386	0.98	638,514	0.88	526,875	0.81	2,445,829	0.92
1994-95	533,894	1.10	806,904	1.02	664,256	0.92	543,086	0.83	2,548,139	0.96
1995-96	545,922	1.13	815,731	1.04	669.621	0.92	549,291	0.84	2,580,565	0.97
1996-97	576,370	1.19	842,846	1.07	693,067	0.96	560,898	0.86	2,673,180	1.01
1997-98	610,330	1.26	873,652	1.11	714,109	0.98	579,058	0.89	2,777,149	1.05
1998-99	633,952	1.31	876.221	1.11	700,108	0.97	575,946	0.88	2,786,228	1.05
1999 2000	648,194	1.34	897,388	1.14	692,397	0.95	585,948	0.90	2,823,928	1.07
2000-01	651,979	1.35	882,837	1.12	671,897	0.93	583,661	0.89	2,790,373	1.05
2001-02	670,094	1.38	890,685	1.13	672,312	0.93	590,783	0.90	2,823,772	1.07
2002-03	699,847	1.45	914,368	1.15	678,736	0.94	612,274	0.94	2,905,226 [sic]	1.10
2003-04	709,680	1.47	914,023	1.16	665,277	0.92	623,114	0.95	2,912,094	1.10

Source: **Western Interstate Commission for Higher Education (1988:13).**

TABLE 2.2 Projected Proportion of United States High School Graduates, by Region, 1986-2004

	West	South/ South Central	North Central	North-east	Total United States
1985-86	0.18	0.30	0.27	0.25	1.00
1986-87	0.18	0.30	0.27	0.24	1.00
1987-88	0.19	0.30	0.27	0.24	1.00
1988-89	0.19	0.31	0.27	0.23	1.00
1989-90	0.19	0.31	0.27	0.23	1.00
1990-91	0.19	0.32	0.27	0.22	1.00
1991-92	0.20	0.32	0.26	0.22	1.00
1992-93	0.20	0.32	0.27	0.22	1.00
1993-94	0.21	0.32	0.26	0.22	1.00
1994-95	0.21	0.32	0.26	0.21	1.00
1995-96	0.21	0.32	0.26	0.21	1.00
1996-97	0.22	0.32	0.26	0.21	1.00
1997-98	0.22	0.31	0.26	0.21	1.00
1998-99	0.23	0.31	0.25	0.21	1.00
1999-2000	0.23	0.32	0.25	0.21	1.00
2000-01	0.23	0.32	0.24	0.21	1.00
2001-02	0.24	0.32	0.24	0.21	1.00
2002-03	0.24	0.31	0.23	0.21	1.00
2003-04	0.24	0.31	0.23	0.21	1.00

Source: Western Interstate Commission for Higher Education (1988:13).

Local-Area Projections

The striking differences in enrollment-based demand projections among the states are mirrored within states by differences among localities. Since teacher labor markets have important local components (as the panel's case studies suggest), it would be useful to be able to produce enrollment projections for local areas. One barrier to local enrollment projections is the reliability of small-population projections that are needed to estimate enrollment in kindergarten and the first grade. Fertility patterns can vary locally, and births and deaths may be reported for an area different from school district boundaries. More important, the smaller the population, the greater the potential influence of hard-to-predict migration on future size and distribution.

The difficulty of making reliable small-area projections is undoubtedly one reason that few school districts appear to make projections beyond the next year, if that far. However, standard enrollment projection techniques would be adequate to give general magnitudes of change in all but the smallest and least stable districts for 5 to 10 years into the future, longer for the secondary level. State departments of education could make a considerable contribution by encouraging school districts to project their enrollment for 5 to 10 years in the future, providing technical guidance in developing projections and coordinating their efforts. Properly done, such projections could be combined in order to approximate likely levels of enrollment-generated demand within teacher labor markets. Of course, estimation of attrition-generated demand and subject-specific demand would require additional projection efforts.

Population Projections—A Proxy for Enrollment Projections

Although the cohort survival method of projecting enrollment is widely used in the education community, population projections reveal some highly interesting trends that could influence the demand for teachers.

Using Census Population Projections to Estimate K-12 Enrollment Demand

At the national level, projections of the population by age provide a very good proxy for enrollment projections, especially if the interest is in total enrollment demand and not just demand for public school teachers. This is particularly true for the population age 5 to 13, which has close to 100 percent attendance, virtually all of it in grades K through 8.

Population projections have the advantage of greater simplicity than enrollment projections, since assumptions about movement from grade to grade or from public to private schools do not have to be made. National projections of the population are updated by the Census Bureau every several years, more often if the underlying assumptions prove incorrect.

The biggest disadvantage to using population projections as a substitute for enrollment projections is that population projections take no account of possible changes in dropout rates, an important element of enrollment projections for the secondary level. Using population projections as a proxy for K-12 enrollment, in particular for grade 9-12 enrollment projections, makes the implicit assumption that dropout rates will remain constant.

Since our present discussion is limited to a general overview of the likely forces of change in the demand for teachers, our remarks on the likely contribution of nationwide enrollment change to teacher demand are based on population projections.

The national projections for the 5- to 17-year-old age group should prove moderately accurate through the year 2000, as shown in Table 2.3. Reliability declines toward the end of the 1990s, when the projections for the age group begin to depend more on projections of births and less on children already born. However, the fertility assumptions used in the projection have been close to actual fertility so far, and there is little reason to expect large changes in the fertility rates in the next few years. The other factor that could lead to the divergence of the actual numbers from those projected—international migration—is unlikely to cause major discrepancies at the national level in the period and at the ages shown in the table. The size of the U.S. population relative even to high levels of migration, and the typical concentration of migration in the early adult years, dampens the effect on the school-age population, at least in the short and medium run. Interstate migration is, of course, not relevant to national projections.

Recent projections by the Census Bureau show 12 percent growth for the school-age population in the United States between the middle 1980s and the end of this century (Table 2.3). The 5- to 17-year-old age group is projected to grow by more than 5 million. However, as a result of past birth patterns, the increase will not be distributed equally across the age group. Until the end of this century, most of the growth will occur at the younger ages and will affect the elementary grades. The number of children age 5 to 13 is projected to increase by nearly 5 million, or 17 percent, by 1999. After 1999 this age group is projected to decline. By contrast, the number of young people age 14 to 17, the secondary school-age group, is projected to decline 12 percent in the 5 years between 1985 and 1990, a reduction of 1.8 million. Thereafter, the number of secondary school-age children will grow slowly but is not expected to regain the 1985 level until 1997. More rapid growth is projected for the early years of the next century.

The projected demographic changes in the school-age population will have opposing potential effects on the nationwide demand for teachers, increasing it at the elementary level and reducing it at the secondary level. There could even be a reduction in the absolute number of secondary school teachers employed over the next few years. However, change in the number of students is only one of the elements in the calculation of demand for teachers. Teacher attrition and pupil-teacher ratios are the other important factors in demand, although pupil-teacher ratios are probably as much dependent on enrollment change as they are an independent factor. The

TABLE 2.3 Projections of the United States School-Age Population to the Year 2000 (in thousands)

	Age 5-13		Age 14-17		Age 5-17	
	Number (000)	Index (1985 = 1.00)	Number (000)	Index (1985 = 1.00)	Number (000)	Index (1985 = 1.00)
1985 (actual)	29,654	1.00	14,731	1.00	44,385	1.00
1986	29,922	1.01	14,588	0.99	44,510	1.00
1987	30,358	1.02	14,237	0.97	44,595	1.00
1988	30,954	1.04	13,662	0.93	44,616	1.01
1989	31,523	1.06	13,160	0.89	44,683	1.01
1990	32,189	1.09	12,950	0.88	45,139	1.02
1991	32,777	1.11	12,964	0.88	45,741	1.03
1992	33,400	1.13	13,087	0.89	46,487	1.05
1993	33,900	1.14	13,260	0.90	47,160	1.06
1994	34,193	1.15	13,714	0.93	47,907	1.08
1995	34,435	1.16	14,082	0.96	48,517	1.09
1996	34,598	1.17	14,513	0.99	49,111	1.11
1997	34,681	1.17	14,848	1.01	49,529	1.12
1998	34,668	1.17	15,027	1.02	49,695	1.12
1999	34,566	1.17	15,214	1.03	49,780	1.12
2000	34,382	1.16	15,381	1.04	49,763	1.12

Source: Bureau of the Census (1984a:43-74).

demand for teachers of specific disciplines, of course, depends on student choice (or changing graduation requirements) as well.

The Changing Demographic Profile of the School-Age Population

Education planners and social observers have devoted considerable attention to the changing demographics of the American population and to projections of large continued changes. The demographic changes referred to are usually changes in ethnic composition and family circumstances, especially increased proportions of children in single-parent families and/or especially increased proportions of children in single-parent families and/or with working mothers. Increased poverty levels among children are often also at issue. The conclusion, sometimes stated explicitly and sometimes left to the reader, is that these changes require urgent social and political attention.

These changes—in the relative size of elementary and secondary school-age populations, in racial/ethnic composition, in income and poverty—are interesting and relevant considerations attendant on the demand for teachers. They do not emerge from enrollment projections, but rather through population projections. We first talk about broad population factors, demographic trends that influence demand. Then we return to the use of enrollment projections in models of teacher supply and demand.

Projecting Changes in Race and Ethnicity. Expected change in the ethnic distribution of school-age children in America is of interest for our discussion only insofar as youngsters of the different categories may be expected to have differential patterns of enrollment in mathematics and science courses or require different strategies of teaching than are used currently. However, the link between ethnic and racial identity and school-related characteristics or needs is not a clear one, particularly over the long run. Unless there is reason to believe that racial and ethnic groups will retain currently observed particular needs over the long run, projecting the racial and ethnic distribution of the school-age population, or of school enrollment, is of little utility for planning curricular or other change for science and mathematics.

Very often, the effect of racial and ethnic change, especially the effect of the projected increases in the proportion of the population of Hispanic and Asian origin, is confounded with the effects of migration—for example, an increase in the number of students with limited ability to speak English or from families with the low educational levels characteristic of Latin America and much of Southeast Asia. This confusion is bemusing in a country that has seen the children and grandchildren of poor, illiterate immigrants from Southern and Eastern Europe—people viewed as forever unassimilable 75 years ago—become thoroughly assimilated Americans.

Changes in racial and ethnic distribution per se may be the least reason for expectations of changed enrollment patterns in science and mathematics or for planning changes in curricular and teaching strategies. However, because of the great public interest in ethnic change, we explore the feasibility of racial and ethnic projections and consider the results of recent projections.

As noted earlier, standard population projections require a base population and assumed rates of fertility, mortality, and migration for each age. The decennial census counts the black population with reasonable accuracy, at least for most public policy analysis purposes. Vital statistics are virtually universally collected by race, including birth and death records. Estimating migration by race is more troublesome, but sufficient data exist to develop estimates of base populations by race between censuses and to develop assumptions for projections. It is feasible to project the black population and the white population, although, as with all subcategory projections, such projections will tend to be somewhat less reliable than projections of the total population. Other racial groups are more difficult to estimate because of small numbers and, in the case of Asians, very high rates of foreign migration.

Projections of nonracial ethnic populations are more difficult. On the whole, ethnic identity has been gathered only sporadically in either the census or in vital statistics. Even were a group to arrive all at once, thereby providing a clear base population, and subsequently maintain accurate birth and death records, the accuracy of any long-range projection would be in doubt because of the likelihood of intermarriage and the lack of an agreed-on definition of ethnic identity for the children. The reasons underlying the questionable feasibility of ethnic projections also raise the question of their meaningfulness for social or educational policy planning.

Because of the interest in the rapid increase in population from Mexico and Central and South America, numerous projections of the Hispanic population have been produced. The Census Bureau first asked respondents to identify themselves as Hispanic or non-Hispanic in the 1980 census. The states with the bulk of the U.S. Hispanic population began to ask Hispanic identity for birth and death certificates around 1980 as well. The data collection efforts since 1980 provide a base population for projecting Hispanics, as well as fertility and mortality rates. Migration estimates can also be made, although with only very modest reliability. Given the importance of migration for determining the size of the Hispanic population, this is a decided disadvantage. More problematic for long-range projection purposes is the lack of a socially agreed-on ethnic identity for the children of marriages in which only one partner is identified as Hispanic.

Results of Racial and Ethnic Projections. The following discussion examines the results of ethnic and racial estimates and projections for the national population by age. In the last two decades, the proportion of American youngsters from non-Hispanic white backgrounds has decreased nationwide and the proportion from Hispanic and from nonwhite backgrounds has risen. This shift is projected to continue, although, as shown in Table 2.4 and described below, the change will be relatively modest at the national level.

In 1970 the census recorded that 13.5 percent of the population age 5 to 17 was black. The proportion rose to 14.8 percent in 1980 and by

TABLE 2.4 Estimates and Projections of the U.S. Population of 5- to 17-Year-Olds, 1970-2000 (in thousands), with Distribution by Race/Ethnic Group

	White (including Hispanic)	Percentage	Black	Percentage	Other[a]	Percentage	Total
1970	44,752	85.2	7,082	13.5	695	1.3	52,529
1980	39,184	82.6	9,009	14.8	1,215	2.6	47,408
1986	36,533	80.9	6,958	15.4	1,653	3.7	45,144
2000 (projected)	38,569	79.0	7,895	16.2	2,351	4.8	48,815

[a] Largely Asian, Pacific Islander, and Native American.

Sources: Bureau of the Census (1987:2; 1989:15).

1986 it was estimated, on the basis of vital records, to be 15.4 percent. The most recent projection is that the population age 5 to 17 will be 16.2 percent black in 2000. The proportion of the 5- to 17-year-old age group reported to be of "other" races, largely Asian, rose from 1.3 percent in 1970 to 2.6 percent in 1980. In 1986 it was estimated to be 3.7 percent and in 2000 it is projected to be 4.8 percent. The proportion white fell from 85.2 percent in 1970 to 82.6 percent in 1980; it was estimated to be 80.9 percent in 1986 and is projected to be 79 percent in 2000. If these projections are correct, the proportion of school-age children who are white would drop by 6 percentage points in the 30-year period from 1970 to 2000. However, data for the white population includes a growing proportion from Hispanic ethnic groups. The shifts in ethnic distribution are not evenly distributed across the country, but are concentrated in certain areas of the country. The proportion Hispanic has increased rapidly in California and other parts of the Southwest, mostly due to immigration, although the Hispanic population also has higher fertility. Asian migration has also been concentrated in a few locations. California and New York City have been the destinations of choice for the majority of Asian immigrants.

The increase in the numbers of immigrants from these two groups is part of an increase in migration to levels not seen in many decades. This increase and the concentration of immigrants in certain areas have heightened the general awareness of shifts in racial and ethnic composition, which are real but less dramatic nationally than might be supposed from the level of popular interest. Providing an adequate education for immigrants and their children is a crucial concern for the schools, especially in areas of heavy influx. It is less clear how important the ethnic shifts, in and of themselves, will be to the schools in the future.

Family Structure and Changes in Poverty Rate for Children. Other shifts in the characteristics of school-age children may be of far more importance to science and mathematics enrollment, as well as to the curriculum that teachers should be prepared to teach, than changes in racial and ethnic distribution. These include changes in family structure, specifically the increase in the proportion of children in single-parent families, and the increase in the proportion of children living in poor families.

Data as of 1985 showed that 16 percent of all white children, 43 percent of all black children, and 40 percent of children of Spanish origin were reported to be living in poor families (Bureau of the Census, 1986:22). The rate of family poverty among all children rose during the 1970s and early 1980s, wiping out gains made in the 1960s. The poverty rate for children was 20.1 percent in 1985, compared with a low of 13.8 percent in 1969 (Bureau of the Census, 1986:22).

During roughly the same period—1970 to 1984—Current Population Reports revealed that the percentage of children living in single-parent families doubled (Bureau of the Census, 1984b:4). In 1984, 22.6 percent of children under 18 were living with one parent, compared with 11.9 percent in 1970 (Bureau of the Census, 1984b:4). In 1984, over half of all black children lived with only one parent, compared with one-sixth of all white children. Among children of Spanish origin, one of every four lived with one parent.

As has been relentlessly demonstrated in innumerable studies, poor children, so many of whom live in one-parent families, are at risk of school failure because of multiple disadvantages, which may include the lack of adequate housing, or any housing at all; frequent moves from school to school; less than sufficient food; inferior medical care; a total lack of dental care; exposure to criminal behavior in deteriorating neighborhoods; and the stress that accompanies the struggle of the adults in the family to survive.

Projections of the proportion of children living with one parent can be made with moderate reliability using current data; the proportion living in poverty requires assumptions about the economy as well and are therefore less easy to construct, or at least less easy to construct with any reliability. The nature of the effect of family poverty and family structure on academic achievement and, more specifically, on the demand for science and mathematics instruction is less well understood. The subject is of great social importance, given the large numbers of children involved, and we hope that the links will be better understood in the future. However, at present, not enough is known to be useful in constructing projections of demand for science and mathematics teachers.

To summarize, there are interesting and relevant trends that emerge from population projections, which should be included in any statistical description of the changing demand for teachers. However, for purposes of projecting demand for precollege science and mathematics teachers, the education community generally finds enrollment-based projections more useful.

Research Areas Related to Student Enrollment

Methods employed in current teacher demand models, specifically the cohort survival methodology used to project enrollments, are relatively reliable. However, for longer-term projections, particularly at the high school level, and for specific subjects within science and mathematics fields, enrollment projections are less reliable due to the impact of changes in the behavior of students, parents, and school systems. The utility of demand models for addressing policy issues concerning science and mathematics education over the long term would be greatly enhanced by the development of more dynamic, behaviorally responsive models. We discuss three types of behavioral responses that need to be understood to develop more useful models of teacher demand. They are determinants of course selection by students, determinants of parental and student preferences for public and private schooling, and changes in dropout rates that can be expected in response to social, economic, and educational changes. We discuss them in the order of their priority as we assess the relative importance of each topic to teacher demand projections and the relative gains that could be expected from research.

As the panel's interim reported stated (National Research Council, 1987c:49), research on the determinants of course selection by students is critical to the development of useful projections for broad subject categories, including science and mathematics, at the high school (and possibly middle school) level. This is an area about which we know very little. Many factors can influence students' choice of courses, including high school graduation requirements, college entrance requirements, government (including federal and state) support for science and mathematics education that motivates schools to encourage enrollment in these subjects, and fashions or tastes on the part of students and their parents and peers for certain subjects.

Given that most current models focus on public school demand (although the National Center for Education Statistics model develops separate public and private school projections), another important area for research concerns the determinants of parental and student preferences for private and public schooling (National Research Council, 1987c:49). Nationwide, private elementary and secondary school enrollment was 11.5 percent of the total in 1980, but had grown to 12.5 percent by fall 1987 (NCES, 1988b:9). Changing preferences for private school enrollment, a topic about which almost nothing is known, can importantly affect public school demand. Particularly in today's educational climate, when private schools are perceived by some parents to offer a more attractive educational environment than public schools, research into the factors that influence the choice of type of school is needed.

One type of response that affects demand projections at the high school level is the dropout rate (National Research Council, 1987c:50). We know a good deal from previous research about why students drop out of school. Work is needed, however, on changes in dropout rates that can be expected in response to a variety of social, economic, and educational changes. For example, the changing ethnic composition of the school-age population in many areas of the country may dramatically affect dropout rates in those areas. Increased high school graduation requirements may increase dropout rates as a side effect of raising educational levels for those who stay in school.

PUPIL-TEACHER RATIOS

Enrollment change does not translate immediately into a corresponding proportional change in the demand for teachers. As a recent RAND report assessing teacher supply and demand explained, "adjustments are made to pupil-teacher ratios to smooth the effects of rapid enrollment changes, to accommodate established school staffing patterns and budgets, and to take into account existing contractual agreements with teachers, in the case of enrollment declines" (Haggstrom et al., 1988:37).

A small change in pupil-teacher ratio can cause a significant change in the projected demand for teachers. Changes in pupil-teacher ratios can be caused by a number of factors at the school, district, or state level: changes in school budgets; staffing patterns, class sizes, or teaching loads; graduation or program requirements; and course offerings. A layering of school, district, and state policies may add to the complexity of factors that change the ratio and the demand for teachers. Even though these factors are complex, they should be identified and discussed briefly as components of teacher demand (Haggstrom et al, 1988:37-38).

Components of Teacher Demand and Related Data

Changes in school budgets can cause changes in staffing practices, class sizes, and teaching loads. The district's allocation of its budget among its many needs for staff, materials, and services affects the pupil-teacher ratio generally as well as specific programs or subjects. In one of the panel's case studies of a school system in a western state, the district filed for bankruptcy following a teachers' strike. This led district officials to make conservative estimates of the number of students expected to be enrolled and hence the number of teachers needed. An overestimated enrollment could cost the district roughly $1,000 per student, it was thought. The tendency to underestimate enrollment and therefore the number of teaching positions has had various effects: raising pupil-teacher ratios, straining teaching loads, or eliminating such support as department chairmanships or resource teachers in disciplines such as mathematics.

Implementation of a school finance formula that changes a district's proportion of local discretionary resources can also affect pupil-teacher ratios. The panel's case studies found that, although some school districts in a southeastern state had enough local discretionary funds to hire additional teachers (part time or full time) in computer science or other subjects, other districts in the state had very little. Local discretionary money—if the district knew the amount far enough in advance—could be used to sign an early contract with a talented candidate for a mathematics resource teacher or an elementary science teacher, for example. Loss of that opportunity could mean leaving the position unfilled. These examples suggest how the budget can directly affect the ability to hire and can substantially affect the pupil-teacher ratio for certain subjects, and ultimately general pupil-teacher ratios.

Changes in staffing patterns, class sizes, or teaching loads may be promulgated by a district rule or policy change, by state policy—or by both, as when a district rule extends beyond a state requirement. These can cause an immediate change in pupil-teacher ratios and in the demand for teachers. A district requirement to employ a full-time guidance counselor in every elementary school, without full additional funding to do so, could strain staffing patterns elsewhere in the school and indirectly push up the pupil-teacher ratio. The Schools and Staffing Survey (SASS), recently initiated by NCES and first fielded in 1988, collects information on staffing patterns, class sizes, and teaching loads. The second SASS survey will be conducted in 1991 and at regular intervals thereafter. As a time series of data becomes available, it will be possible to monitor changes in these variables over time. SASS includes a teacher demand and shortage questionnaire for public school districts and private schools, as well as a school administrator questionnaire for public school principals and private school heads. It also includes a teacher questionnaire and a teacher follow-up survey: a one-year follow-up survey of the sample teachers who have left teaching and some who have remained. This ongoing, integrated survey effort has been designed to provide the most comprehensive data on teacher demand and supply available to date. Although we mention the survey frequently as a potential data source, as with all new surveys, the extent to which it will live up to its expectation cannot be known until policy makers and the research community have used the data in a variety of analyses.

Course requirements in science and mathematics for graduation (usually established by the state) clearly affect pupil-teacher ratios. As an example of such policies and responses that can cause changes in pupil-teacher ratios, most states and school districts have increased their graduation requirements since 1980 (NCES, 1988c), often adding additional science and mathematics course requirements. The Center for Policy Research in Education (CPRE) reports that since 1983, 42 states have added course credit requirements in science, mathematics, or both (CPRE, 1989). NCES has found, through its Fast Response Survey System, that the state requirements are often exceeded by the requirements of individual districts (personal communication, M. Papageorgiou, NCES, June 7, 1989). Unless more teachers are assigned or hired to teach science and mathematics, the pupil-teacher ratio for these subject areas clearly increases.

Some nationally collected data on high school graduation requirements are available on a regular basis. The Education Commission of the States (ECS) and the Council of Chief State School Officers publish information periodically on state-mandated high school graduation requirements. They track mathematics and science as general categories, however, listing only the number of courses or years of science and mathematics that are required for graduation.

SASS includes an item in the teacher demand and shortage questionnaire for public school districts on high school graduation requirements, by subject (physical and biological sciences, mathematics/computer science). It asks for changes in these requirements between 1987 and 1988. Future SASS results will reveal changes in requirements over longer time spans.

Course offerings and enrollments also influence pupil-teacher ratios. Whereas requirements clearly help determine what courses high school students take—and the demand for teachers of those subjects—an important constraint is whether the required courses are actually offered. For example, only a few schools offer a complete range of college-preparatory mathematics and science courses; a physics course might be offered only every other year. And very few students are enrolled in the most advanced courses.

Course offerings and enrollments by school, school system, and state emerge as an important variable. Course offering data could serve as an excellent indicator of science and mathematics demand by either students or state requirement. Course offerings and enrollments would contribute toward indicating change in demand over time, by size of school and school system, by other relevant school district characteristics, and by state, especially when state graduation requirements for science and mathematics courses have changed. Course offerings and enrollment would also permit analysis of the degree of school response in terms of teacher assignment. The extent to which teachers need to teach more than a single subject could be noted. Some data related to course offerings and enrollments by school system and school are being gathered nationally.

The 1985-86 National Survey of Science and Mathematics Education (Weiss, 1987) provides the most recent comprehensive data on course offerings. Data on course offerings are also included in (1) the longitudinal study High School and Beyond, (2) in the National Educational Longitudinal Study of 1988 (NELS:88) (for middle schools and junior high schools), and (3) for students age 9, 13, and 17 in the National Assessment of Educational Progress (NAEP) in those years when science or mathematics achievement are assessed. While data from these surveys are disaggregated by specific subject areas, they are conducted infrequently.

The data on course offerings and course enrollments are "plagued with inconsistencies," according to a recent report on elementary and secondary science education (Office of Technology Assessment, 1988:42). Course titles often are not a reliable basis for comparisons among schools, states, or years. Moreover, some advanced courses are offered not by the high school but by the community college, and there are no national data available on this practice. National data do not show how often a physics course is given. Nor do we know how many sections of a given course are taught. These data may change, as well, from year to year in a single school.

NELS:88 asks middle schools and junior high schools for data on numerous courses and whether they are offered. In a more detailed format, questions on course offerings are included in the SASS teacher questionnaire. Given a probability sample of teachers by fields (as is the case for the teacher sample), it should be possible to estimate the prevalence of course offerings and trends in course offerings (including science and mathematics offerings) at national and regional levels.

Finally, course offerings could serve as a basis for drawing samples to test for varied working conditions, recruitment patterns, and range of initial assignments possible among school systems. Vacancies matched to schools classified by offerings might indicate conditions of low retention. The SASS questionnaire of local education agencies (LEAs) asks district administrators for the total number of positions that are either vacant, filled by a substitute, or withdrawn for lack of a suitable candidate; this total is disaggregated by subject.

Research on Determinants of Pupil-Teacher Ratios

A closer analysis of often interrelated factors that influence pupil-teacher ratios is a rich area for further research. As noted earlier, in most models pupil-teacher ratios are estimated in a relatively arbitrary way. But numerous factors operate and interact to cause changes in pupil-teacher ratios for science and mathematics subjects and in general. And we suspect that certain types of dynamics in teacher markets (e.g., declining enrollments or increased school budget) may be associated with declining pupil-teacher ratios. Other conditions (e.g., surging enrollments, budget cutbacks) are associated with rising pupil-teacher ratios. Since that ratio is so critical to an assessment of the demand for teachers, research on its determinants is needed.

The factors that can change pupil-teacher ratios affect adjustments over both short-term and long-term periods, although short-term adjustments differ from longer-term ones (National Research Council, 1987c:50). For example, a shortage situation may result in a marked increase in pupil-teacher ratios until the school system has had time to implement responses, such as extended recruitment or hiring teacher aides.

A common practice in projecting the demand for teachers is to project the increase in enrollment and to divide it by the current pupil-teacher ratio to calculate the number of additional teachers needed to provide for the enrollment increase (National Education Association, 1987f:14). But an assumption that using the current, general pupil-teacher ratio reflects accurately the number of teachers is too simplistic; enrollment changes affect pupil-teacher ratios in more indirect ways. For example, research suggests that, when enrollments decline, teacher unions may be willing to forgo salary increases to keep current teachers employed (Cavin et al., 1985; noted in Haggstrom et al., 1988:42). Despite enrollment declines, school boards may decide not to lay off staff in science and mathematics if they have had difficulty hiring them in the past, or if they feel it will be hard to find qualified teachers of certain subjects in the future (Prowda and Grissmer, 1986:12).

Moreover, supply-demand projections for precollege science and mathematics teachers will be far more meaningful if both enrollments and pupil-teacher ratios are disaggregated by subject area. In Connecticut, for example, secondary enrollments have been declining, but the demand for secondary science and mathematics teachers is steady or may increase because of increased graduation requirements in these subjects, coupled with decisions made (influenced by budgetary considerations) to decrease class sizes (Prowda and Grissmer, 1986:12).

There is reason to believe that pupil-teacher ratio is a dependent as well as an independent factor in the creation of demand. In periods of enrollment growth and teacher or financial shortages, the ratio (or class size) can be increased. When demand slackens, if all "surplus" teachers are not let go, then the ratio drops. A history of these coping responses would be extremely useful in developing better assumptions about pupil-teacher ratios for demand models than the assumption that the current, general pupil-teacher ratio is adequate. The development of demand models for science and mathematics teachers would require similar information on the history of class size responses to teacher shortages or surpluses within these disciplines, a formidable undertaking but necessary for really well-defined models.

In sum, we continue to recommend, as we did in our interim report, that research be conducted on the determinants of pupil-teacher ratios, including research on adjustment lags as enrollments change and on how changes in demand for courses contribute to changes in these ratios and in the demand for teachers of science and mathematics.

TEACHER ATTRITION RATES

The third major element in the construction of teacher demand models is the rate at which teachers leave their jobs. It should be noted that teacher attrition is largely a supply phenomenon, reflecting the decisions of individual teachers. In Chapters 3 and 4 we treat attrition as a supply variable, but here it is natural to think of it as resulting in a demand for new teachers.

One part of the leaving rate—retirement—is fairly easy to model, if data on the teaching force by age are available and if something is known about the typical ages at which teachers retire. Rates of attrition for other reasons are much less easy to determine. Some of this information exists in school records, but it must be gathered and put into forms usable by those developing models. In aggregating attrition data gathered from school districts it is important to avoid double counting, since what is attrition to one district might be a new hire to another.

The Connecticut model of teacher supply and demand revealed how important age is in estimating attrition rates (Prowda and Grissmer, 1986): "We have noted high early career attrition rates, low mid-career attrition, and high attrition around 60 and 65 years old" (p.1). It is likely that attrition rates change over time, reflecting the numbers of teachers hired in a given year or period (high attrition rates may be observed soon after). A RAND study of teacher attrition found a similar U-shaped pattern of attrition in Illinois, Michigan, New York, and Utah (Grissmer and Kirby, 1987:36-38). Higher attrition rates were found among newly hired teachers than among other groups, including those eligible to retire.

In addition to reflecting age or years of experience, calculating attrition for mathematics and science teachers requires gathering the necessary data by specific field, a considerably more demanding task than obtaining the data for the entire teaching force. There are studies that have gathered discipline-specific data (Murnane and Olsen, 1989a, 1990b; Grissmer and Kirby, 1987), but these data have not often, if at all, been used in the development of teacher demand models (or supply models in which they would also be of use). As part of the fifth follow-up of participants in the National Longitudinal Study of 1972, completed in 1986, a Teaching Supplement Questionnaire was sent to sample members who were teachers, former teachers, and those who had been trained to teach but had not taught. Information from the survey included detailed professional and personal histories that could be used for analyses of attrition patterns during the early part of their careers, from 1977 to 1986. Heyns (1988) analyzed the data, but results were not reported by field of discipline. The SASS questionnaires were designed to provide national data on teacher attrition by field. The SASS public school questionnaire asks for the number of teachers, by field, who left in the previous year and their destinations. However, due to low response rates for these items, researchers have to depend on the SASS teacher follow-up survey of former teachers (conducted in 1988-89) for estimates of attrition by field. NCES staff is exploring alternative ways of obtaining better attrition data. The follow-up survey, which also asks for the destinations of leavers, is expected to provide national attrition rates by field.

Attrition has important demographic elements, in part because so much of it is caused by retirement. Since something is known of the demographic profile of the teaching force, it is possible to estimate the likely general trend of attrition in the future, which is almost certain to be on the rise. In much of the country, low rates of new hires over the past 10 years and reductions in force, with the newest teachers being the ones let go, have left a relatively senior task force. Table 2.5 shows an upward trend from 1976 to 1986 in the proportion of current teachers who are age 40 and over (from 34.6 to 51.3 percent) and who are age 50 and over (from 15.5 to 21.2 percent). A rise in the rate of retirements will, of course, increase attrition. In addition, evidence noted earlier points to particularly high levels of attrition in the early years of teachers' careers.

In order to forecast attrition adequately, more information is needed not only on the distribution of teachers by age, but also by disciplinary area and level of preparation, and on the current attrition levels within those categories. It would also be useful to have a better understanding

TABLE 2.5 Age Distribution of U.S. Public School Teachers, 1961-1986

	1961	1966	1971	1976	1981	1986
	Years					
Mean	42	39	38	36	39	41
Median	41	36	35	33	37	40
Under age 30	[a]	33.9%	37.1%	37.1%	18.7%	11.0%
Age 30-39	[a]	22.8	22.8	28.3	38.8	37.7
Age 40-49	[a]	17.5	17.8	19.1	23.1	30.1
Age 50 and over	[a]	25.8	22.3	15.5	19.4	21.2

[a] Subgroup data not available.

Source: National Education Association (1987e:73).

of why attrition differs for the different categories, if it does, so that more reliable assumptions can be developed for projection models. These issues are discussed in more detail in Chapters 3 and 4, which focus on supply.

SUMMARY

The demand for teachers, as we have indicated, depends on enrollment changes, both generally and in mathematics and science courses. The demand for teachers also depends on changes in pupil-teacher ratios, caused by changes in staffing patterns, class size, teaching loads, course requirements, and course offerings in mathematics and science. In addition, a school district's demand for new science and mathematics teachers in a given hiring season also depends on the number of vacancies in those subjects. The number of vacancies results not only from the creation of new positions, but also from teacher attrition, a component of supply.

In general, the panel considers the data available for projecting demand to be more adequate than data for projecting supply. The task of projecting enrollment-driven demand for science and mathematics teachers is relatively straightforward.

There is a small number of significant gaps, however, in data related to demand, and the panel recommends collecting data to fill these gaps. Forecasting the demand for science and mathematics teachers particularly could be improved by better data on the following variables:

Course-taking behavior in high school. Data on state-mandated high school course requirements, collected regularly over time and by science or mathematics subject, could suggest changes in demand for teachers of various types or levels of courses. School district requirements often exceed state requirements, but with both state and district data we could begin to trace how changes in course requirements stimulate changes in demand for secondary science and mathematics teachers.

Changes in course offerings in science and mathematics. Changes in course offerings can change the demand for science and mathematics teachers and can identify the need for teachers with special skills, for example, ability to teach advanced placement physics.

Enrollment changes disaggregated into science and mathematics course enrollments. Better data on this aspect of course-taking behavior, in conjunction with changes in course requirements, would strengthen the demand component of projection models.

Data on attrition for reasons other than retirement by field. Attrition by retirement is relatively well known. For other types of attrition, further analysis of the NLS-72 follow-up of teachers and former teachers should provide more insight into patterns of attrition during the early years of a teaching career (Heyns, 1988). The best source for obtaining new nonretirement attrition data will be SASS, which has recently experimented with questions on attrition by field, although the item response rate was low. High priority should be given to collecting attrition data because they are essential to both demand and supply models. The panel urges that NCES redesign the SASS questions on attrition and subject them to a thorough pilot test before using them. When combined with other SASS data on teachers, the attrition data could help answer questions such as: Among mathematics teachers and science teachers who leave earlier in their careers, how many had taught advanced courses? Introductory-level courses? In high school or middle school? The demand created by such patterns will thus be better known, and a closer fit may be possible in filling the demand.

In addition, research is suggested on the behavioral factors that influence the demand for teachers, particularly teachers of science and mathematics in the higher grades, for use in development of improved models for longer-term projections. Among the research areas noted are the behavioral determinants behind course selection, factors that influence dropout rates, influences on parents' choice of public versus private schools, and the relationship between demand for certain courses and pupil-teacher ratios.

3
Determining Supply: Individual and District Activities

Each year between 5 and 10 percent of the nation's public school teachers leave the profession. Some leave permanently; some leave temporarily; and they leave for a variety of reasons—to take a different job, to pursue further education, to start a family, etc. What this means, however, is that every year between 100,000 and 200,000 replacements are needed to fill those vacancies, although the actual number is not known. From where do these teachers come?

The panel attempted to answer that question, particularly for secondary school science and mathematics teachers, by looking at three sources of evidence: (1) state and national models of teacher supply and demand; (2) in-depth case studies of classroom teacher recruitment in a number of school districts; and (3) insights obtained through a conference on professional personnel systems in large school districts.

In this chapter we first discuss what constitutes supply—continuing and new science and mathematics teachers—and their incentives and decisions about teaching along stages in their career paths. We then look at supply from the district viewpoint, from which widely varied policies for recruiting, screening, and selecting teachers cause variations in the adequacy and the quality of the supply of teachers available to different districts.

THE COMPONENTS OF SUPPLY

The supply of teachers for the coming school year is a relationship between the number of qualified individuals who would be willing to teach and such incentives as the salaries, benefits, retirement programs, working conditions offered by school districts, and other alternative career opportunities. Ideally, it would be desirable to have a behavioral model of supply that would take into account the interaction and interdependence of a wide range of variables and could help answer such questions as how many teachers can be expected to quit in response to a change in retirement policy, or how many former teachers can be expected to reenter if salaries are raised by a certain amount.

Policy makers frequently ask questions about the likely impacts of various education policy actions and socioeconomic forces on prospective teacher supply and demand. To address such questions requires a capacity to project supply and demand under varying assumptions about future circumstances. In turn, this capability requires the development of models that are both behavioral and dynamic. By this we mean models that capture relationships between variables in the environment and the behavior of actors in the educational system, and in particular capture relationships between changes in circumstances and subsequent changes in the numbers and kinds of people interested in obtaining teaching positions or in the numbers and kinds of teachers demanded by school systems.

Before such models can be developed, additional research on the relation between incentives and supply and between variables in the environment and supply, as well as additional data to support the models, will be needed. The national and state models examined by the panel are projection models based on extrapolations of current conditions or historical trends, although some use refinements such as age and field-specific attrition rates in projections of continuing teachers and consideration of a broader range of new supply sources. In practice, these models try to estimate the number who will be available from each of the two major components of supply: continuing teachers—teachers who are teaching this year and will continue to teach next year in the same location—and new entrants. There is a continuous flow of teachers into and out of the teaching force, as shown in Figure 3.1. This diagram can apply to the nation, a state, a school district, or to special groups of schools such as rural or inner city schools, or to special types of teachers such as science teachers, mathematics teachers, or minority teachers.

Continuing Teachers

The most important element of teacher supply during a given year is the retention of people returning from the prior year. To obtain that component of teacher supply, we need to know the attrition between the two years. However, the attrition rate is a complex function depending on the various incentives that cause teachers to retire, to move to another school, or to leave teaching for other careers including homemaking. In practice, the method typically used in current models involves making an assumption about attrition rates, sometimes adjusted for trend and sometimes not. For

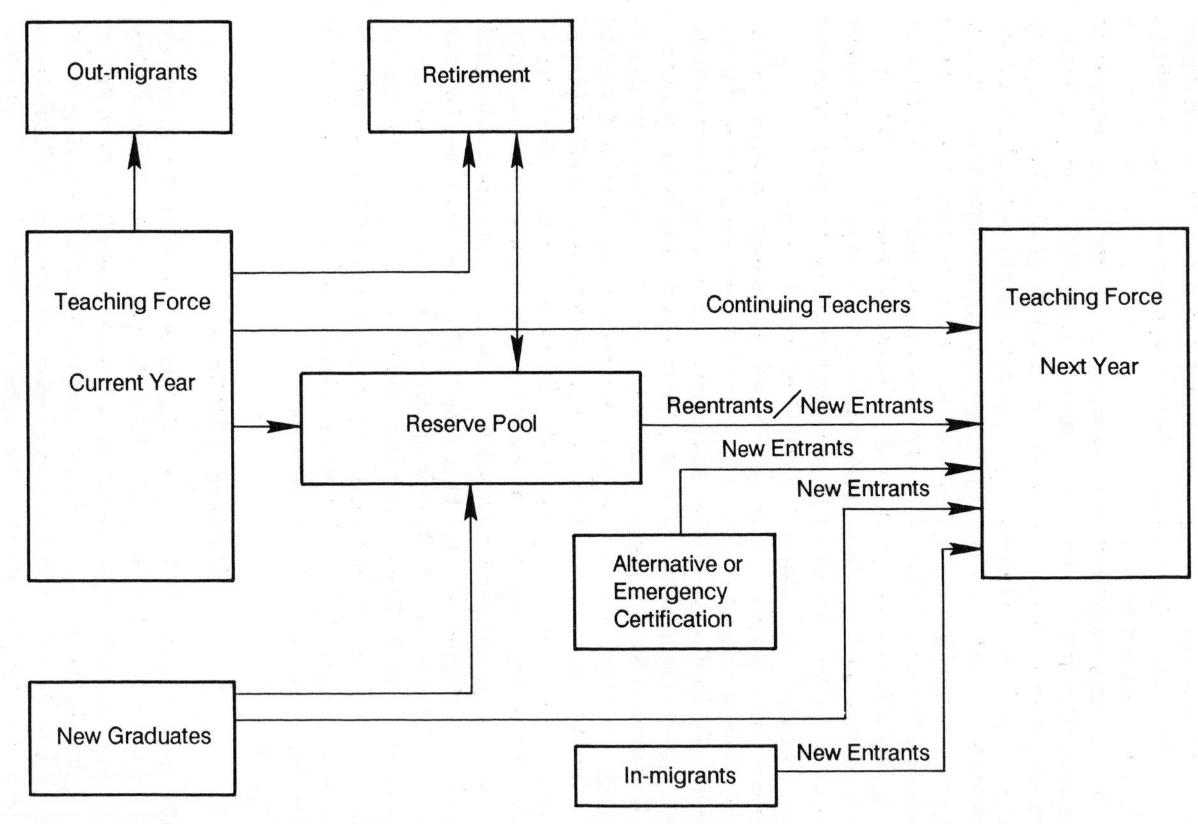

FIGURE 3.1 Supply: Components of the teacher force.

many years the model used by the National Center for Education Statistics for the intermediate set of projections used an attrition rate of 6 percent based on survey data that were collected by NCES in 1968 (Metz and Fleischman, 1974). In 1987 NCES used an estimated turnover rate of 7.5 percent for elementary teachers and 6.5 percent for secondary teachers (CES, 1987a:46). The results of the Schools and Staffing Survey (SASS) should provide a basis for more accurate attrition rates.

In state models, the supply due to retention can be estimated in a more satisfactory way, since states generally have information on the attrition rate for teachers in the state for the prior year and assume that the rate will be the same in the current year. However, attrition levels vary from state to state and over time. Attrition for states is different from national attrition, since the former includes teachers who move to other states and continue in teaching. Some states also have the information to compute more refined attrition rates, such as for different age groups of teachers and for different subject fields. (Evidence from these states shows that as the teaching stock ages, the average attrition rate will change.) Interestingly, attrition rates in these states for teachers of science and mathematics are not noticeably different from rates of teacher attrition in other fields. Tables 3.1 and 3.2 show retention rates for public school teachers in the states of Illinois and New York for mathematics, science, and the total for all subjects. (As we point out later, the lack of difference in rates by subject may be due to the influences of general enrollment declines during the early 1980s.)

TABLE 3.1 Retention Rates for Illinois Public School Teachers 1977-1984 (Percentage Retained in Consecutive Years)

		Secondary Grades (9-12)		
Year	Primary Grades (PreK-8)	Mathematics	Science	All Subjects
1977-1978	90.5	91.7	92.1	90.8
1978-1979	90.4	91.7	92.3	90.2
1979-1980	91.6	92.3	90.5	91.9
1980-1981	91.8	92.5	92.5	92.1
1981-1982	92.6	93.0	93.8	93.3
1982-1983	92.5	94.6	94.5	93.0
1983-1984	93.5	n.a.	n.a.	93.4

Note: Data are for downstate schools only (i.e., all school districts except the Chicago Public Schools). This table shows retention rates of teachers; however, the source publication shows attrition rates, i.e., 100 = the retention rate.

Source: Illinois State Board of Education (1983: Tables 2 and 3; 1985b: Table 8).

TABLE 3.2 Age-Specific Retention Rates of New York Public Secondary School Teachers (grades 7-12) in 1984 (Percentage Retained from 1983)

Age	Mathematics	Science	All Secondary Subjects
Under 35	90.6	89.9	89.2
35-39	94.1	95.6	94.4
40-44	94.9	95.6	94.8
45-49	95.8	95.8	95.2
50-54	91.9	93.0	92.0
55-59	84.2	84.0	83.5
60 and over	73.2	69.1	70.2
Total all ages	92.4	92.7	91.6

Source: New York State Education Department (1985a).

New Entrants

The more difficult part of modeling teacher supply consists of predicting the potential willingness of people who were not teaching last year to enter the teaching force. In Figure 3.1 we have labeled all sources of teacher supply other than continuing teachers as "new entrants" or "reentrants." Major categories under the heading of new entrants include newly certified persons, persons with previous teaching experience and certification (i.e., reentrants—people who come from the so-called reserve pool of teachers), persons hired through some alternative or emergency certification procedure, and in-migrants.

The major categories can be broken down into yet finer components. Newly certified persons may be either newly certified graduates of teacher training programs or newly certified graduates with other majors. Experienced teachers may have been on leave or layoff, they may have entered other careers (including homemaking); they may have been teaching as substitutes; they may have resigned for long-term health reasons; or they may be in-migrants. In-migrants are teachers who were teaching last year, but not in the particular jurisdiction or subject field for which the supply is being estimated. In some states virtually any college graduate, with or without teaching certification or experience, can be counted in the supply of new entrants; these states permit certification on the basis of testing, permit hiring on an emergency certification basis, or use an apprentice teaching program.

In the first phases of our deliberations, we discovered that the major proportion of new hires each year did not come from new college graduates, but rather from the corps of experienced returning teachers. Although the percentages varied across subject areas, level, and location, in general it was found that less than half of new hires were new college graduates (National Research Council, 1987c:27). For example, Table 3.3 shows that the proportion of new hires who were new college graduates was less than 30 percent in each of six types of urban-suburban-rural districts. The National Education Association's (NEA) surveys of American public school teachers found a decline over the years in the proportion of new entrants who came directly from college (NEA, 1987e:24). From the data provided, the percentage of new hires who had been in college the previous year can be computed to be 67 percent in 1966, decreasing to 17 percent in 1986. These findings are important in light of the fact that the supply-demand model used by the National Center for Education Statistics until 1987 based its estimates of teacher shortage on the assumption that all new hires would be new college graduates. Following the publication of the panel's interim report, NCES discontinued this practice.

WHAT INFLUENCES AN INDIVIDUAL TO TEACH?

From an examination of teacher supply, the panel has concluded that the answer to the question "Who will teach science and mathematics in the nation's schools?" is heavily influenced by the incentives offered to teachers, former teachers, and potential teachers. This conclusion follows from the results of a long history of studies showing that the supply of skilled labor for particular occupations is sensitive to financial incentives (see, for example, Harris, 1949; Arrow and Capron, 1959; Freeman, 1971). This section summarizes what is known about the role that particular incentives play in the career decisions of teachers, former teachers, and potential teachers. We discuss in subsequent sections the extent to which the important incentives play a role in teacher supply and demand models, or could play a role in improved models.

Although this report is concerned with science and mathematics teachers, most of the literature on teacher supply does not distinguish between these teachers and other elementary and secondary school teachers. Consequently, we must look to the broader literature for evidence on incentives and teachers' responses to them.

TABLE 3.3 First-Year Teachers as a Percentage of New Teacher Hires by Type of District and Subject Area, for New York State Public Schools, 1985-86

Subject Area	New York City	Other Large Cities [a]	Other City Districts	New York City Suburban Districts	Other Suburban Districts	Non-Suburban Districts
Common branch (grades 1-6)	23.7 %	18.2 %	19.9 %	16.4 %	15.1 %	25.8 %
Total elementary	23.6	20.7	22.0	18.2	19.0	27.1
Secondary						
English	22.9	16.3	14.6	11.3	22.9	24.0
Foreign language	24.0	26.7	26.3	16.9	25.4	31.0
Mathematics	25.5	16.7	19.7	19.0	23.4	34.4
Earth science	25.8	15.4	46.2	36.4	37.3	45.9
General science	25.1	23.1	23.5	34.5	45.0	30.8
Biology	29.3	33.3	36.7	21.6	26.4	42.7
Chemistry	35.0	16.7	40.0	27.9	19.4	55.8
Physics	21.7	11.1	50.0	22.7	26.9	55.9
Social studies	28.2	24.1	30.0	12.9	25.0	31.4
Occ. ed.	21.3	29.8	25.7	14.7	35.0	36.3
Total secondary	24.6	20.5	24.9	16.7	26.1	34.1
Total teachers	22.7	20.5	22.4	16.5	21.7	29.6

[a] Cities over 125,000 population.

Source: New York State Education Department (1988).

Before turning to this evidence, we want to make clear that this discussion should not be interpreted as implying that financial incentives are the only factors that influence teachers' and potential teachers' career decisions, or even that they are the most important influences. Teachers enter teaching for a variety of reasons—to work with children, to experience the satisfaction of helping others, to have a schedule similar to their own children's schedule. Other reasons for entering teaching were given by some of the newly hired teachers interviewed in the panel's case studies. They entered teaching because of particular experiences they had in the past—teaching opportunities during college that were rewarding or an outstanding individual high school teacher who served as a role model, for example. Teachers also leave teaching for a variety of reasons—to pursue another occupation, to follow a spouse whose job has been relocated, to engage in full time childrearing. For most teachers and potential teachers, a moderate change in salary, say $2,000 to $4,000, probably does not influence the decision about whether to enter teaching or how long to stay in teaching. However, a critical question is whether such a moderate-sized salary change would influence the career decisions of enough college graduates to have a marked influence on supply. That question is addressed here.

It is also important to keep in mind the unit of analysis that provides the focus for particular studies of the determinants of teacher supply. For example, a number of studies report that recruitment efforts by individual school districts have been successful in expanding the quantity and quality of applicants for teaching positions. Presumably the reason is that these efforts have made particular school districts seem especially attractive to a significant portion of the pool of potential teachers. It does not follow, however, that active recruitment policies by all school districts would improve the quantity or quality of science and mathematics teachers in our schools. Instead, these policies are likely to influence only the distribution of the available supply of teachers among different districts. The implication of this example is that, when evaluating the evidence on responses to incentives, it is important to consider the extent to which the incentives alter the quantity and quality of the pool of science and mathematics teachers available to the nation's schools, or whether they influence only the distribution of the available supply among districts.

This section focusing on individuals is organized according to what might be called the steps in the pipeline that place teachers in schools:

1. College students' decisions about occupational preparation;
2. The decision about whether to enter teaching;
3. Teachers' decisions about how long to stay in teaching;
4. Former teachers' decisions about whether to return to teaching;

5. Teachers' decisions about moving from one state to another; and
6. Teachers' decisions about when to retire.

College Students' Occupational Preparation Decisions

Over the last 15 years, the percentage of American college students training to become teachers has declined precipitously. One indicator of this is the proportion of graduating seniors majoring in education. This proportion has fallen from 22 percent in 1971-72 to 9 percent in 1985-86 (National Center for Education Statistics, 1988b:210). This indicator is suspect, however, because an increasing proportion of college students training to teach also major in a particular discipline, for example, mathematics or biology. As a result, the trend in the number of education majors may provide misleading information about the trend in the number of college students preparing to teach. Unfortunately, no reliable national data exist on the number of individuals obtaining teacher certification each year. This makes it necessary to turn to individual states for information on the number of new certificants. Table 3.4, which provides information on the number of individuals obtaining certification in New York and North Carolina in selected years between 1974 and 1985, illustrates the dramatic decline in the number of individuals obtaining teacher certification in these states. In each state the number of new certificants in 1985 was less than half of the number of new certificants in the mid-1970s.

There are two related reasons why the number of college students training to teach declined dramatically over the last 15 years. The first is the decline in the number of teaching positions available for newly certified teachers—a response to enrollment declines. For example, the number of new teachers (that is, teachers without previous teaching experience) hired by public school districts in Michigan declined from more than 6,000 in 1973 to fewer than 700 in 1984. In addition, many beginning teachers lost their jobs as fiscally strapped school districts reduced staff in response to enrollment declines. Since the probability of obtaining a teaching position is a critical factor influencing college students' decisions about whether to train to teach, the decline in this probability was an important factor contributing to the decline in the proportion of college students preparing to teach.

A second factor was the decline in teaching salaries relative to salaries offered by business and industry. As depicted in Figure 3.2, teaching salaries fell relative to salaries in business and industry during the late 1970s. Thus, as papers by Manski (1987) and Zarkin (1985) have shown, the combination of the decline in the probability of obtaining a teaching position and the decline in the competitiveness of teaching salaries were strong signals to college students to pursue occupations other than teaching.

TABLE 3.4 Number of People Obtaining Teacher Certification in New York and North Carolina, 1974-1985

Year	New York	North Carolina
1974	34,770	
1975		6,538
1976	24,039	6,413
1977		5,673
1978		5,105
1979		4,684
1980	16,348	3,852
1981		3,145
1982		3,095
1983		3,071
1984	17,275	2,997
1985	16,002	2,830

Sources: New York State Education Department (1988); Murnane and Schwinden (1989:9, Figure 1).

The Decision to Enter Teaching

One of the surprising facts about the operation of the teacher labor market is that one-third to one-half of college graduates who obtain teacher certification never teach—or at least do not teach in the state where they obtain certification.[1] One explanation is that teacher certification has traditionally been relatively easy to obtain in most states and, as a result, many college students obtain certification even though they have little interest in teaching. A second explanation is that the decline in the number of teaching vacancies during the 1970s left many newly certified graduates without job offers in teaching.

The fact that a large proportion of graduates certified to teach do not teach raises the interesting question of who enters teaching and who does not. Recent work by Murnane and Schwinden (1989) indicates that the answer varies across subject specialties and race. They found that the National Teachers Examination (NTE) scores of white certificants trained

[1] National data on who is certified to teach are not available, and state-level data provide no information on certificants who leave the state.

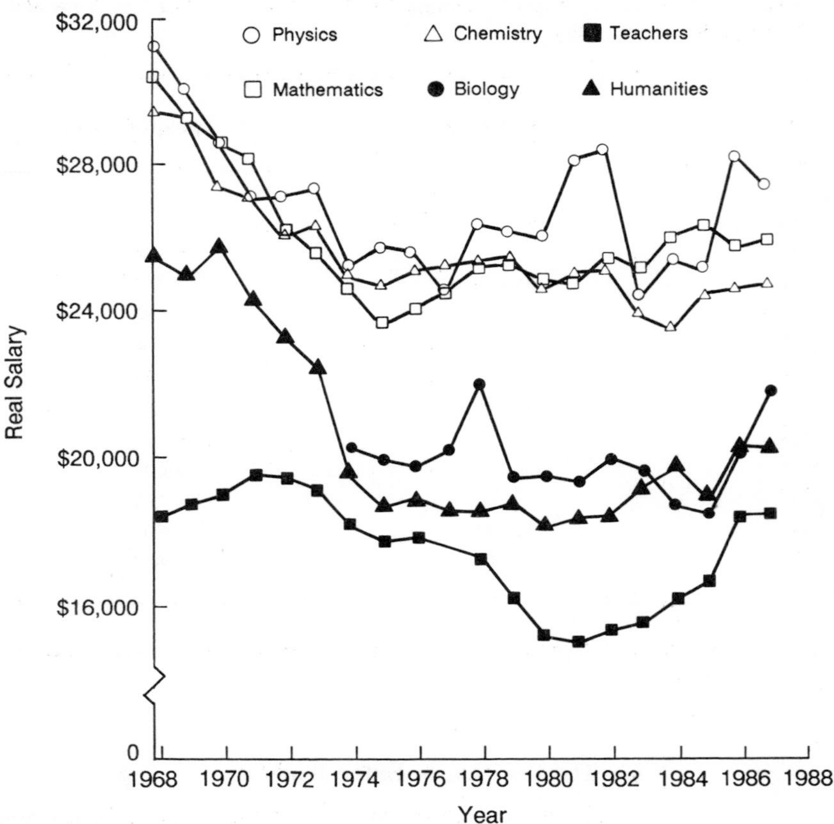

FIGURE 3.2 Average starting salaries for teachers and for business and industry ($1987) by college graduates' field of college major. Sources: College Placement Council (1988); National Education Association (1987a).

in chemistry, mathematics, and English were important predictors of the probability of entry into teaching. The entry probabilities for white certificants in these areas with scores at the 90th percentile were 10-17 percentage points higher than the entry probabilities for white certificants in these areas with scores at the 10th percentile. (NTE scores made the most difference in the probability of entry for certificants specializing in mathematics.) For white certificants with other subject specialties, the NTE score was not an important predictor of the probability of entry.

The likely explanation for this pattern concerns the opportunity cost of becoming a teacher—that is, what one gives up if one decides to teach. As presented in Figure 3.2, between 1968 and 1987 starting salaries in business and industry for college graduates trained in chemistry or mathematics were considerably higher on average than starting salaries for graduates trained in biology or the humanities. If graduates in chemistry and mathematics with high scores on the NTE (a standardized test whose scores are positively correlated with scores on other standardized tests, such as the Scholastic Aptitude Test and the Graduate Record Exam) are more likely to receive offers of high-paying jobs in business and industry than are graduates with low scores, this would explain the negative relationship between NTE scores and probability of entry into teaching for certificants in these fields. (This explanation also implies that job offers in business and industry for graduates majoring in English are more attractive than job offers for graduates majoring in history—an assumption that we cannot test.) Although we have not demonstrated that salary differences by field affect career choice, the key point here is that the evidence supports the proposition that a college graduate's choice of occupation depends on relative salaries. Murnane and Schwinden's study found that there was no negative relationship between NTE score and probability of entry into teaching for black college graduates certified to teach. The reason may be that black graduates faced less attractive job opportunities in business and industry than did white graduates, or that they were more place-bound than were white graduates.

The implications that one draws from the negative relationship between NTE score and the probability of entry into teaching for white college graduates trained in chemistry and mathematics depend on one's assessment of the relationship between NTE score and teaching effectiveness. If NTE score were a strong predictor of teaching effectiveness, the results would imply that the profession of teaching is losing a high proportion of the most promising potential teachers. However, in a review of the history of NTE scores, Haney, Madaus, and Kreitzer (1987) find little if any correlation between teachers' NTE scores and other measures of teacher effectiveness, such as supervisor's ratings. What does follow from this evidence is that higher teaching salaries may be a necessary condition for recruiting teachers who have the skills to do well on standardized tests (e.g., to score above the 10th percentile) and who have subject specialties such as science and mathematics that are highly rewarded in business and industry.

Where to Teach

One of the unique characteristics of public education in the United States is that hiring is done quite independently by 15,000 school districts that establish their own salary schedules (usually through bargaining with local teachers' unions) and design their own recruitment and screening procedures. As a result of differences in salaries, working conditions, and recruiting practices, there is significant variation in the ability of local school districts to attract and retain skilled science and mathematics teachers. As a result, one must be careful when using the term *teacher supply.* It is quite possible that districts that pay high salaries, have good working conditions, and have aggressive recruitment practices may be successful in attracting skilled teachers at the same time that districts without high salaries, and districts in which working conditions are difficult, cannot attract skilled teachers at all. Indeed, this is the inference that the panel has drawn from the case studies on district hiring practices and from the conference with personnel directors of large, urban school districts.

Unfortunately, relatively little is known about the incentives that are important determinants of school districts' ability to recruit skilled teachers. There is little documentation about the extent to which school districts' recruiting and screening strategies influence their ability to hire skilled teachers, as opposed to less skilled teachers. Typically, even with low salaries, difficult working conditions, and poor recruitment practices, a district can find adults to stand in front of classrooms; however, they are unlikely to be skilled teachers. In other words, the adjustment mechanism concerns quality, not quantity. Consequently, studies that examine only whether districts have applicants for teaching positions, without paying close attention to the skills of the applicants, do not provide reliable information about the influences of school district salaries, working conditions, and recruiting practices on the ability to staff the schools with skilled teachers.

How Long to Stay in Teaching

One element of teacher supply that has an important influence on the demand for new teachers is the length of time that teachers already in the schools stay in teaching. As Grissmer and Kirby (1987) have pointed out, even a small change in the percentage of teachers who leave teaching from one year to the next (the attrition rate) has a dramatic change on the demand for new teachers. As explained in the panel's interim report, the national teacher demand model used by the National Center for Education Statistics assumes implicitly that: the attrition rate is constant over time; the attrition rate is not influenced by changes in teacher salaries; and the attrition rate does not vary among subject specialties. Recent research examining the factors that influence the length of time that individual teachers stay in teaching call all of these assumptions into question.

There is a long history of studies showing that attrition rates follow a U-shaped distribution (see Grissmer and Kirby, 1987, for a list of references). Young, inexperienced teachers tend to have very high attrition rates—often as high as 20 percent in the first year. The probability that a teacher leaves teaching declines with experience. Attrition rates are very low for teachers with more than five years of experience. Finally, attrition rates begin to climb again as teachers near retirement age.

A series of studies employing data from Michigan, North Carolina, and Colorado (Murnane and Olsen, 1989a, 1989b, 1990; Murnane et al., 1988, 1989) have demonstrated that teaching salaries are an important determinant of the length of time that teachers stay in teaching. The evidence implies that a $1,000 annual increase in salary (in 1987 dollars) is associated with an increase of one to two years in the median length of time that teachers stay in teaching.[2]

These studies also show that high school teachers tend to stay in teaching for shorter durations than elementary school teachers do—a pattern present in all three states. The studies also find some differences in the career paths of secondary school teachers with different subject specialties. For example, Figure 3.3, which is based on a sample of North Carolina teachers who began their careers in the late 1970s, shows that chemistry and physics teachers tended to leave teaching sooner than did secondary school teachers with other subject specialties. It is important to point out, however, that the sizes of the differences in career paths by subject specialty vary across sample and time period. This is illustrated in Table 3.5, which displays predicted median first spell lengths in teaching for teachers with different subject specialties. As explained in Murnane and Olsen (1990), the predicted survival functions for teachers with different subject specialties are based on a model in which length of first spell was modeled to be a function of age at entry, gender, subject specialty, NTE score, annual salary expressed in 1987 dollars, and a dummy variable for the district in which the teacher started his or her career. Notice that in the two states in which it is possible to differentiate chemistry-physics teachers from biology teachers, the latter group has a higher median first spell length. This suggests that, in discussing teacher supply, it is important not to treat science teachers as a homogeneous group. This quantitative evidence is also supported by the comments of several personnel directors who were interviewed as part of the panel's mini case studies. The frequent comment was that there were plenty of strong candidates for biology positions but, in some districts, a shortage of strong applicants for teaching some other sciences, especially physics.

A number of recent studies have examined attrition rates by subject specialty using an approach different from the research described above, which is based on the analysis of longitudinal data on teachers' careers. These other studies develop estimates of age-specific or experience-specific

[2]The methodology Murnane and Olsen use controls for time-invariant district-specific characteristics, even unmeasured ones. In effect, their model includes a dummy variable for every district.

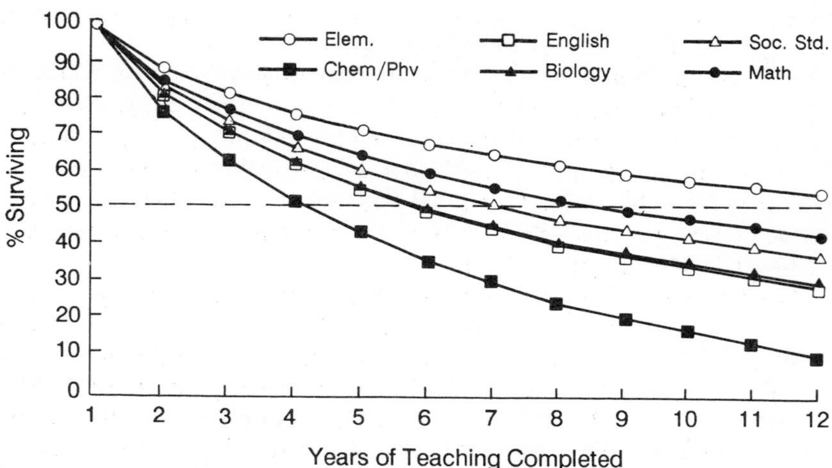

FIGURE 3.3 Predicted survival functions by subject specialty, for a sample of teachers in North Carolina. Source: Murnane and Olsen (1990).

attrition by comparing the rosters of teachers employed in a state or in a school district in two consecutive years (Grissmer and Kirby, 1987). The analytic strategy is to calculate the proportion of teachers with particular characteristics (for example, mathematics teachers between the ages of 30 and 34) who were teaching in the first year, but not in the second. An advantage of this approach is that it is possible to calculate attrition rates using very recent data.

One significant research puzzle is that the studies using longitudinal data tend to find greater differences in attrition rates by subject area than do the studies comparing cross-sections of teachers for two consecutive years. The likely explanation concerns the timing of the data. The research comparing cross-sections tends to be based on data from the mid-1980s, when declining school enrollments led to involuntary attrition in many school districts. These involuntary quits, which tend to be based on seniority, may mask differences in voluntary quits that are sensitive to opportunity cost. The studies using longitudinal data tend to be based on the careers of teachers who started to teach in the 1970s. These teachers may have acquired enough seniority by the time enrollment declines set in to be relatively free from involuntary layoffs. Thus, the attrition patterns observed in the studies based on longitudinal data may be less influenced by the consequences of enrollment declines. If this explanation is correct, then the studies based on longitudinal data—studies that show significant differences in first spell lengths by subject specialty—may predict attrition patterns in the 1990s better than the studies based on more recent data.

TABLE 3.5 Predicted Median First Spell Length (in Years) of Teaching: Samples of Teachers from Three States

State	North Carolina	Michigan	Colorado
Period Sample Began Teaching	1975-79	1972-75	1979, 1982
Sample Size	(8,462)	(7,785)	(1,377)
Teaching specialty			
Elementary school	13.5	16.4	6.6
Mathematics	7.9	7.4	4.4
Social studies	6.6	7.6	3.4
English	5.7	7.3	3.1
Biology	5.6	9.6	6.0
Chemistry-physics	4.1	4.9	[a]

[a] Chemistry-physics teachers cannot be distinguished from biology teachers in the Colorado data.

Source: Murnane and Olsen (1990).

The reason is that secondary school student enrollments (age 14-17) will increase in most parts of the country in the 1990s; as a result, involuntary layoffs will be rare, and in some regions there may be a teacher shortage.

Whether to Return to Teaching

Until recently, the national teacher supply and demand model used by NCES assumed that newly minted college graduates provide the only source of teacher supply available to fill new vacancies. Recent data from several states suggests that this assumption may be seriously out of line with the current situation. For example, 75 percent of the individuals newly hired by Connecticut school districts for the 1986-87 school year had prior teaching experience and were returning to teaching after a career interruption (Connecticut State Department of Education, 1987). The analogous number for New York State for the 1984-85 school year is 70 percent (New York State Education Department, 1987). These data raise the question of whether the "reserve pool," consisting of individuals fully certified to teach but not currently teaching, will be an important source of supply in the years ahead. The question is difficult to answer since there is little systematic knowledge about the size of the reserve pool, its composition, and the factors that influence members of the reserve pool to return to teaching.

Studies based on data from North Carolina, Michigan, and Colorado throw some light on this question by examining whether teachers who left teaching within the first five years after entry[3] returned to the classroom after a career interruption (Murnane and Olsen, 1989b; Murnane et al., 1988, 1989). The evidence indicates that approximately one-third of elementary school teachers return to the classroom after a career interruption. The return rate for secondary school teachers varied by field from 10 to 30 percent, with teachers of mathematics, chemistry, and physics having the lowest return rates. Among high school teachers in Michigan, teachers of chemistry and physics were the least likely to return to the classroom. In North Carolina, secondary mathematics teachers were the least likely to return. This evidence is quite consistent with the notion that former teachers' career decisions are sensitive to relative salaries. Those former teachers with subject specialties that paid relatively high salaries in business and industry (shown in Figure 3.4) were much less likely to return to teaching than teachers with subject specialties that paid lower salaries in business and industry. Thus, the limited evidence currently available suggests that the reserve pool is less likely to be a significant source of supply of chemistry, physics, and mathematics teachers in the future than it will be a source of teachers in other fields, especially elementary education. But to the panel's knowledge, the size of the reserve pool is unknown.

Whether to Move to a Different State

As a result of demographic trends that include migration and differential fertility rates, some parts of the country experience shortages of teachers while other parts of the country do not. One logical solution to this problem is migration of teachers from areas of oversupply to areas of excess demand. Clearly some mobility exists, but the design of state pension systems for teachers is a significant deterrent to relocation. As Bernard

[3]The data bases Murnane and his colleagues examined include 12 years of longitudinal information on teachers' careers. In examining the return rate of teachers, they focused on teachers who ended a first spell of teaching within five years, in order to provide a significant period of time for teachers to return after a career interruption.

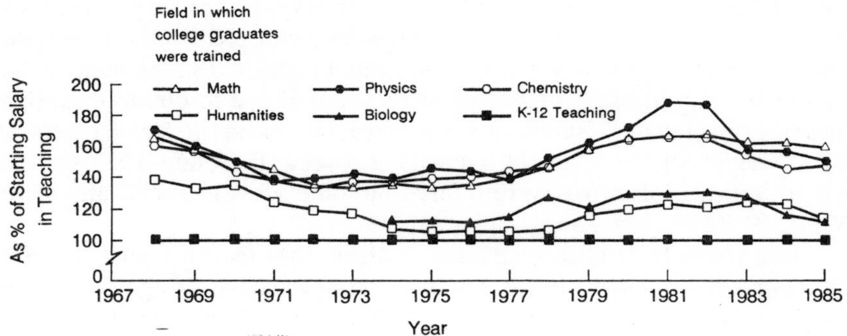

FIGURE 3.4 **Starting salaries in industry as a percentage of starting salaries in teaching.** Sources: **College Placement Council (1988); National Educational Association (1987a).**

Jump (1986) has explained, the design of most state pension systems is such that a teacher who moves from one state to another ends up with a significantly smaller pension than do teachers who continue to teach in the same state.

A second deterrent is the character of state certification systems. While a significant amount of reciprocity among states exists, it is not always possible for a teacher certified in one state to obtain certification automatically in another state. In fact, recent efforts to stiffen certification requirements in individual states may reduce the amount of reciprocity.

The designs of pension systems and certification systems influence the balance between teacher demand and supply because they influence the extent to which a critical equilibrating mechanism, movement of teachers from areas of teacher surplus to areas of teacher shortage, operates. Changes in pension portability rules, a policy recently discussed by the National Governors' Association (1988), may have a significant influence on the rate of interstate mobility of teachers, and consequently, may influence the balance of teacher supply and demand.

When to Retire

The conditions that influence the timing of retirement for teachers are similar to those for individuals in other occupations. Teachers tend to retire when they feel financially able to do so; when incentives for early retirement are considered worthwhile; and when increases in benefits for continuing another year or two are minimal (Taylor, 1986).

The most common provision for maximum retirement benefits occurs at 60 years of age and 30 years of service. States provide for early retirement below these maximum levels with penalties for either years of service, age, or a combination of both (Ginsberg et al., 1989). States vary from these provisions, with some having easier and some stricter provisions. Five states require only 25 years of service for maximum benefit; 10 states require that teachers be employed 30 years and be 65 years of age to obtain maximum retirement payment; 7 states provide maximum payment for teachers 55 years of age with 25 years of service. Between 1979 and 1982, special early retirement incentives were offered by some school systems and states (Wood, 1982).

Until recently, studies seemed to indicate that teachers tend to retire at the earliest age allowed; this would mean approximately age 60 (Taylor, 1986). In South Carolina, a study asked teachers if they would retire if there were a hypothetical change in the provisions of the retirement law. The current provisions permit retirement after 30 years of service, or at age 65, or at age 60 with a penalty. The change would permit retirement after 25 years of service or at age 55. The responses showed this change would increase separations by 5.5 percent above those under the current law, and a change to retire at any age after 25 years of service would increase separations by 16.6 percent above those under the current law, or 10.5 percent above the level with age 55 (Ginsberg et al., 1989).

However, teachers in some states deviate from this pattern. For example, in 1983, approximately 20 percent of the teachers employed in Maryland were eligible to retire. Only about half of these teachers had retired by 1988. Their decisions were undoubtedly heavily influenced by the state's policy with respect to provision of the employee's share of health benefits and cost-of-living salary adjustments—a policy that did not have an age cap. This combination of incentives to stay beyond retirement age is unusual—41 states have an age cap on provision of health benefits and 20 place an age cap on cost-of-living benefits. Policies about provision of health and cost-of-living benefits obviously influence decisions to retire or delay retirement.

Many factors other than age can affect the decision to retire. When a large wage increase is expected, some teachers will delay retirement to obtain a higher average salary. A change in options for health plans can hasten or impede retirement, especially when retirees can not change plans after separation. When a large cost-of-living adjustment in retirement benefits is expected, retirements can increase. During a period of large cost-of-living increases, such as in the late 1970s, retirements can decrease because of worry about having sufficient retirement funds. Since these factors affect all teachers, the combined influence may lead to a bunching of retirements.

Although separation rates for retirement are known by the state retirement systems (Kotlikoff and Smith, 1983), this information has not been used either for strengthening supply-demand statistics related to science and mathematics teachers or for research. To obtain a better understanding of the retirement decision, research is needed on the relationships among separation rates, individual reasons for early or late retirement, external variables, and incentives and disincentives for remaining in the teaching force or retiring.

In short, although we know the effect of separation rates on states, we also need information about their effect on individual school districts and on the teaching force by level and by discipline or subject taught.

Conclusions

The basic theme of this section is that the number of teachers in a given field willing to work in a given location depends on a number of incentives, including the availability of teaching positions, salaries, opportunity cost salaries, working conditions, certification rules, and pension rules. Changes in any of these incentives will influence supply.

For two reasons, however, the panel is skeptical about the feasibility of improving projection models by incorporating the influences of these incentives. First, there is a great deal of variation across school districts and across states in salaries and working conditions, and it is difficult to incorporate this variation in an aggregate model. Second, while a number of studies show that teachers' career decisions depend on salaries, the coefficients indicating the sizes of the impacts are quite sensitive to sample definitions and estimation techniques. In other words, research is not sufficiently developed to provide reliable estimates of the response coefficients that could be included in projection models. For these reasons, the panel does not recommend at this time the development of projection models of teacher supply and demand that include responses of teachers and potential teachers to changes in incentives; however, the panel does urge support for the development of better behavioral models to measure the sensitivity of teacher supply to incentives (see Chapter 6).

The panel also recommends that resources be devoted to monitoring trends in the levels of key incentives that influence teacher supply. Timely information about changes in the proportion of newly certified teachers who obtain teaching positions, the competitiveness of teacher salaries relative to opportunity cost salaries, the amount of reciprocity in certification across states, and the portability of teacher pensions may provide policy makers with early warnings about likely changes in the supply of teachers.

A number of research issues related to the decision about when to retire and its effect on teacher supply need investigation. Separation rates for retirement are known by state retirement systems and could be used in research relating these rates to individual reasons for separations, to external shock variables, and to incentives and disincentives for retention and retirement. Related research issues include the effect of separation rates on individual school districts and on the teaching force for different fields of study.

HOW DOES A DISTRICT MESH SUPPLY WITH DEMAND?

School district policies, practices, and constraints exert considerable influence on an individual's decision to apply for a position, take a position, remain in a position, or leave. In teacher supply and demand models, school districts are treated as black boxes. No models incorporate information on school districts' recruitment, screening, and hiring processes in the structure used to generate predictions of teacher supply and demand. As a result, the design of the models implicitly assumes that variation in these practices do not have a marked impact on the ability of individual school districts to attract skilled math and science teachers. While the variation in practices does not matter in using models to project the supply of teachers in a state or in the country, it does suggest that the models do not provide reliable information about the supply of skilled teachers available to individual school districts. Recent case study evidence (Berry, 1984; Wise et al., 1987) has called this assumption into question by pointing out that there is considerable variation among school districts in recruiting, screening, and hiring practices and that these practices may have a marked influence on the ability of school districts to hire skilled teachers.

For this reason, the panel wanted to learn more about the school district practices that affect the supply and demand of science and mathematics teachers. We were concerned with the flow of teachers through and within the school districts. We also hoped to obtain insights concerning variables at the school or district level that affect demand and supply. To pursue this goal, the panel first commissioned the development of detailed case studies of the recruitment, selection, and retention of science and mathematics teachers in six districts, which varied in size, student clientele, enrollment trend, wealth, and location. Second, the staff, in supporting the panel's activities, conducted supplementary case studies focusing on supply and demand issues affecting science and mathematics teachers in 24 districts. These supplementary studies used telephone conversations with personnel directors and a follow-up mail survey to collect more detailed information about the hiring of science and mathematics teachers in each district. Third, the panel convened a conference of the personnel directors of seven of the nation's largest public school districts, which represent over 5 percent of the public school enrollment in the United States. Topics for discussion in the day-and-a-half-long meeting included: effective recruiting strategies, experiences with the reserve pool, recruitment during the school year, supply-demand models, and information system design and use. (See Appendix A for more detail.)

In both the case studies and the conference of personnel directors, differences among the districts were as apparent as the commonalities. Observations made by the school district officials during the course of these three activities are noted throughout the report.

This section summarizes the lessons the panel has learned from these activities concerning the variation in school district practices and the perceptions of how practices influence school districts' success in recruiting skilled math and science teachers. The section is organized by topics corresponding to the following elements of the hiring process: determining needs, soliciting applicants, screening applicants, and making offers. The section emphasizes school district practices because they provided the focus for the case studies and the conference with the personnel directors. However, it is critical to keep in mind that determination of who teaches in the schools depends not only on these practices, but also on applicants' responses to these practices.

Determining Needs

Knowing how many new teachers of each subject at each grade level will be needed in the coming year is a critical first step in planning a hiring strategy. Yet, for many districts, it is extremely difficult to collect this information in a timely fashion. Some reasons are detailed below.

Uncertainty About Student Enrollments

Student enrollments are the primary determinant of the demand for teachers. Based on comments from the personnel directors of seven large school districts, it appears that projecting future enrollments accurately is difficult to do, especially in districts experiencing significant in-migration or out-migration. Since there are many such school districts, the panel infers that many districts do not have reasonably accurate projections of enrollments for a given year until the students actually appear in September. When finances preclude flexibility in the ratio of the number of students to teachers—a situation present in most of the districts included in the case studies—teachers cannot be hired in anticipation of enrollment increases. This inability to offer firm contracts to strong applicants in late spring, when many applicants desire commitments of employment, hinders many districts' efforts to hire skilled teachers.

Uncertainty About Budget for the Next School Year

Another related problem is that in the spring when hiring takes place, districts may not know the budget for the next school year. According to union contracts, this forces the district to inform large numbers of teachers in March that they will not be employed the following year. Then when the budget is assured, they find that some of the people who receive notices have found other work. Thus, the district must look for new people.

Internal Transfer Queues

In many districts, contracts with teachers specify a formal procedure under which teaching vacancies are made available to teachers already employed by the district, before they can be filled by a newly hired applicant. Completing the steps of the internal transfer process often takes several months. Until the process is completed, the personnel office cannot be sure of the identity of the school in which a vacancy will ultimately be present, or even of the teaching specialties that will be needed.

Delays in Reporting Resignations

For a number of reasons, teachers may delay reporting that they plan to resign their positions. One reason is that some contracts specify that teachers employed by the district on the date on which a new contract is signed are eligible for certain fringe benefits included in the new contract, such as improvement in health benefits. As a result, teachers wait until a new contract is signed, which often runs into the summer months, before resigning. Another reason for late resignations is that some school principals will ask teachers who intend to resign to withhold formal notification so as to subvert the internal transfer process. Principals do this to gain control over who fills the vacancy. One consequence of this practice, however, is a wave of resignations in late summer, when it is difficult to find qualified applicants.

Attrition During the School Year

While most suburban districts and smaller-sized districts tend to concentrate their recruiting on finding strong applicants in late spring to fill vacancies expected for the following September, many urban districts hire teachers throughout the year to fill unanticipated vacancies resulting from teacher resignations and unexpected enrollment growth. In fact, the personnel directors from several urban districts reported that as many as half of the teachers they hire are asked to start teaching during the school year, rather than in September.

Soliciting Applicants

School district personnel directors use a variety of strategies to recruit applicants for teaching positions. These include recruiting at nearby colleges and universities, relying on informal networks of information about individuals not currently teaching who are interested in returning to the classroom, and, in one urban district, recruiting graduate students to teach part time. These are a few examples of the many strategies that personnel directors described as effective for finding applicants.

The need to recruit varies greatly among school districts, and large differences are observed in ratios of applicants to vacancies reported by districts in the same labor market area. For example, in the Washington, D.C., metropolitan area, the District of Columbia has great difficulty attracting applicants; it reports about three applicants for each teaching job. By contrast, suburban Montgomery County reports a 13 to 1 ratio; and Prince George's County 8 to 1 (Sanchez, 1989). The wide array of recruiting strategies employed reflects such differences in the ability to attract applicants.

School systems may advertise and make trips to job fairs or colleges where they have successfully recruited in the past. If personnel officials feel there is a particular shortage, special early offers may be made. At times, to eliminate a particular shortage, special incentives, such as a bonus, may be offered. What seems clear from personnel administrators is that many school systems are searching nationally, or at least beyond their local or state borders, for persons in similar fields. One year the quest may be for science and mathematics teachers; another year it will be for reading teachers; still another year, it may be for early childhood teachers. Lately a widespread need has been for special education teachers and for teachers of the same ethnic backgrounds as those of the students in the district. According to the personnel directors of large school districts who shared their experiences with the panel, recruiting generally was restricted to known sources, because experience had taught recruiters that persons unfamiliar with the climate, housing costs, student populations, or culture were unlikely to remain in their systems.

While there was enormous variation in the way personnel directors found applicants, some patterns emerged. First, almost all personnel directors indicated there was no shortage of qualified applicants for teaching positions in science or mathematics at this time. (Most districts did report shortages of minority applicants and applicants for special education positions.) Exceptions were a few cases of an inadequate supply of applicants to teach physics. Several respondents commented that the supply of qualified applicants for each vacancy in biology was considerably greater than the ratio of qualified applicants to number of vacancies in chemistry or physics. Second, most respondents indicated that a large percentage of their applicants were individuals with previous teaching experience.

While it is hazardous to make inferences about patterns in recruiting strategies from an unrepresentative sample, the following patterns seemed to be present in the survey information. School districts that paid relatively low salaries, particularly districts in rural areas, relied especially heavily on attracting applicants who had grown up in the area and were eager to return home. Districts that paid high salaries and offered attractive working conditions found that many of their applicants were teachers currently employed in nearby districts. Urban districts that needed large numbers of new teachers each year were more likely than other districts to engage in national recruiting strategies. While some personnel directors of large districts indicated that they did find national recruiting worthwhile, they were quick to point out that hiring applicants from regions of the country with very different climates often led to very high turnover rates. As a result, they had learned to concentrate their recruiting efforts on geographical areas that had supplied a relatively large number of applicants in the past, and had found it particularly fruitful to recruit in areas in which teachers were being laid off as a result of declining enrollments and budget cutbacks.

Screening Applicants

The strategies used to screen applicants for teaching positions, including who does the screening and how it is done, varied considerably from district to district. Our discussions with school districts indicated that recruitment of new teachers by large school systems with diverse student populations was often hindered by the fact that recruiters could not specify the school to which the applicant would be assigned. Many persons would find such a school system desirable if they could teach in a given section of the school system or in a specified school. Since recruiters could not make those promises or could not make those promises soon enough in the recruitment period, candidates were lost to the school system. In other school systems of various sizes, however, the school and the position was able to be specified early in the screening process.

To provide a sense of the variation in screening applicants, we describe the screening practices in two school districts in the Northeast—the subject of a recent Harvard University doctoral dissertation (Shivers, 1989). (The districts are not identified because a commitment of confidentiality was made to the study districts.) The first district is an ethnically and socioeconomically diverse community adjacent to a central city, and has a long-standing reputation of providing excellent education. The second is an urban district with a history of budget problems and difficulties in raising the achievement of a clientele containing a large proportion of low-income and minority children.

In the first district, the screening process is extremely decentralized. The central personnel office weeds out unpromising applicants and passes on to school principals a list of promising candidates. In the words of the acting superintendent, who had been personnel director for a great many years (as quoted in Shivers, 1989):

> In a nutshell, our aim here is to use the central staff to do a paper screening of the candidates, to do some initial interviewing, and then to forward as quickly as possible as many reasonable candidates as possible to the building principal or the curriculum coordinator [department chair] at the secondary level . . . and then to let them do the selection. That is to say that we sort of send out a group of people with the Good Housekeeping Seal of Approval that are a general batch. And from that general batch the principal should choose.

At the high school level, the school principal delegates to the department chair authority to choose among applicants. The logic underlying this practice is that chairs are responsible for the quality of instruction offered in their departments and for evaluating teachers. Consequently, they should be responsible for hiring the teachers who will provide the instruction to students. Using open-ended questions, chairs interview each candidate sent from the central personnel office. They also call references. They are not obliged to choose among the candidates sent to them. If none seems satisfactory, they can ask the personnel director to find other candidates. They can also use their own informal networks, such as professional associations and experiences with substitute teachers, to find candidates.

Chairs indicated that they do not attempt to hire teachers fitting one mold. Rather they look for candidates who know their subjects, demonstrate evidence of teaching skills, and also do something special, so as to maximize the probability that they will appeal to a subset of the school's diverse student population. One indication of the diversity that is sought is that in one year during the early 1970s, when 115 teachers were newly hired, there were 87 different graduate backgrounds, 25 states, and 7 countries represented (Shivers, 1989).

One attribute that chairs do seek in applicants is some (but not too much) teaching experience. One chair summarized this priority by stating that she did not want a "person with a B.A. degree and no experience . . . . This is too complicated a school to take children to teach children. If I had my druthers they [great candidates] would have had two years experience somewhere else so they would have made their really bad [teaching] mistakes somewhere else" (Shivers, 1989). Although the superintendent is formally responsible for hiring teachers, in practice the authority is delegated to principals, who in turn delegate it to chairs. In fact, after the chairs complete their interviewing and choose the candidates they want to hire, they convey their choice to the winning candidate and also call the candidates who are not chosen. Thus, the decentralization goes beyond advice-giving. In practice, the department chairs choose the teachers.

In the second district studied in this doctoral dissertation, the screening process is markedly different. Principals and building-level department chairs have only a minor role. The superintendent and central office assistants play the major role in determining who will be hired. As in the first district, the process begins with a central office screening of the credentials of applicants. Potentially acceptable candidates are asked to come for interviews. It is at the interview stage at which the process in this district is so different from the one-on-one interviews between candidates and building-level department chairs that were used in the first district. Shivers (1989) describes the interview process used in the second district as follows:

> All new teacher candidates are asked to gather at the same time at the . . . high school gymnasium to be interviewed. Interview panels, which are put together by the personnel director and by department heads, include three to six interviewers [typically including] the appropriate district-level department chair, a secondary school principal, a building-level department chair and another teacher from the department, and one or two central office administrators . . . .
>
> Teachers are called one by one to face a panel of interviewers who are seated at a table on the gymnasium floor, out of earshot of the candidates waiting in the stands . . . .
>
> Lists of questions are prepared beforehand by the respective district-level department chairs. Before the interviews, panel members choose from the list the five or six questions that their panel will ask. The same questions in the same order must be asked of each candidate . . . . During the interview, the panel members rate each answer as positive, negative or neutral . . . .
>
> Panel members are not permitted to respond to the candidate's answers, and no follow-up questions are permitted . . . . The strict procedure for interviewing was developed in response to concerns voiced by the affirmative action office . . . that there be no preferential treatment of candidates for teaching jobs.
>
> After all candidates have been interviewed, the respective teams rate their candidates as highly recommended, recommended, or not recommended. Generally, they do so by consensus. The ratings are sent to the assistant superintendent for personnel who then checks references of recommended candidates . . . . The superintendent or assistant superintendent may interview top candidates after they have been recommended by the panels. At this stage the superintendent will make the final selection.

Clearly, the screening processes in the two districts studied by Shivers (1989) are extremely different. The experience of applying for a teaching position differs greatly in the two districts. What cannot be known from Shivers' work is whether the screening practices influence who ultimately is hired. This could occur because the department chairs in the first district look for skills that are different from those sought by the superintendent in the second district. It could also occur because the screening processes influence the size and quality of the applicant pool or the rate at which potentially effective teachers accept job offers. Informal networks of college placement officers, college faculties, or students may have a great deal of information about how districts screen applicants and the effects on applicants. Unfortunately, there has been virtually no systematic research about potential applicants' responses to differences in recruiting and screening practices.

The case studies commissioned by the panel revealed considerable variation in screening practices among districts—both in the degree of centralization and in the relative roles played by paper credentials, test scores on standardized written tests (which some districts administer as part of the screening process), and interviews. While the case studies do not provide a basis for describing the distribution of screening practices among the nation's 15,000 school districts, they do verify that practices vary enormously. They also raise the question of the extent to which variation in these practices influences the ability of school districts to hire teachers who are effective in teaching math and science to students.

Making Offers

The case studies revealed enormous variation in the types of offers made to candidates whom school districts would like to employ. Dimensions of the variation include timing, specificity concerning the nature of the position, and salary.

Timing

Personnel officers in some school districts, especially well-financed, growing districts, are authorized to offer binding contracts to strong candidates before the exact number and composition of vacancies are known. Several personnel directors suggested that this practice facilitates their recruitment efforts by allowing them to recruit aggressively in colleges and universities during the spring months and to sign up promising candidates before other districts had ascertained the number and nature of their vacancies. Other personnel directors told about the other side of the coin, losing promising candidates because their districts prohibited offering contracts until firm information on vacancies was available, which often took until late summer.

Specificity

In the first of the two districts that were studied by Shivers (1989), being offered a contract meant that the candidate knew a great deal about the position: the school building, the subjects to be taught, the grade levels, and the name of the department chair. In the second district, a teaching contract meant only a commitment to salary. Not only did the newly hired teacher not know the building or the classes to be taught, but also the new teacher did not know the date on which this information would be available.

The case studies commissioned by the panel indicated that the examples described by Shivers are not particularly unusual. Typically, in smaller districts, candidates are told more about the details of their teaching position than in larger districts. However, in some large districts staff at the school site play a significant role in the screening process and, in these districts, candidates are often hired to teach in a particular school.

Salary

In the 24 districts included in the panel's supplementary case studies, the starting salary for a candidate with a B.A. and no teaching experience ranged from $14,420 to $26, 061. The starting salary for a teacher with an M.A. and the maximum amount of experience that the district rewarded ranged from $25,956 to $47,941. Some of the differences in salary scales were responses to differences across communities in the cost of living. However, the comments of the large district personnel officers indicated that the salaries they could offer played a significant role in the ability of school districts to attract a strong applicant pool and to capture the most capable candidates from the pool.

Another important aspect of hiring practices revealed by the case studies is that the formal salary schedule in many districts does not totally determine the salary offered to a newly hired teacher. For example, Dade County offers a $1,000 signing bonus (paid in the first check of the second contract year) to new hires in shortage areas. The first of the two districts Shivers studied sometimes convinces especially strong candidates in shortage fields to sign contracts by giving credit in terms of steps on the salary schedule for practice teaching and for experience outside teaching. Current contracts in Boston and Rochester include specific language allowing the district to do the same thing.

Not all variation from the salary schedule involves pay increases. The second of the two districts Shivers studied frequently hires teachers as permanent substitutes instead of as regular contract teachers. This provides an annual savings of $9,000 to the fiscally troubled district. In this district, attempts are made to attract strong candidates in science and mathematics by offering them regular teaching contracts instead of positions as full-time substitutes. Unfortunately, no information is available on the impact of this practice on the district's ability to attract strong candidates.

Who is Hired

The case studies revealed enormous variation in the practices school districts use to recruit, screen, and hire teachers. It is not possible from these studies to determine the extent to which the variation in practices influences the ability of districts to attract strong candidates. In fact, the case studies revealed that there is not even a common definition of a strong candidate. The remarks of personnel directors suggest that districts' constraints and practices do matter. For example, the notes contain many comments about losing candidates either because salaries were not competitive or because the district could not make a firm contractual offer, while another district could. Some districts can hire early, and those that can have a better choice of candidates. Maintaining close ties with a local teaching credential program also helps bring strong candidates.

Moreover, there is the distinct possibility that school district practices matter less in the late 1980s than they will in the 1990s. The reason concerns the potential change in the overall balance between teacher supply and demand. With the exception of a few fiscally constrained urban districts, most districts included in the case studies reported an adequate number of qualified candidates for each vacancy in mathematics and science. Most districts also reported that many applicants were experienced teachers, and that they filled a large proportion of vacancies with experienced teachers.

This reliance on older or experienced applicants raises the question of whether the responses of personnel directors in 1987 and 1988 provide reliable predictions of the adequacy of the supply of qualified math and science teachers in the years ahead. In the late 1980s, the demand for new secondary school teachers is relatively low because high school enrollments are not growing. At the same time, the reserve pool of individuals certified to teach but not currently teaching appears to be quite large, in part because it contains many individuals from the large cohorts born at the tail end of the post-World-War-II baby boom. The 1990s will be characterized by growing demand for science and mathematics teachers, both because of modestly growing secondary school enrollments and of increasing numbers of resignations from an aging teaching force. At the same time, the size of the reserve pool may decline, both because the number of individuals in the 30-40 age group will be smaller and because the anticipated general labor shortage will bring about more competition for all skilled workers. In an environment characterized by a shortage of qualified applicants for teaching positions, school district practices in recruiting, screening, and hiring teachers may have a considerable impact on the distribution of qualified teachers across school districts.

Conclusions

The evidence on school district hiring practices has two implications for understanding teacher supply and demand. First, as the demand for new hires increases in the 1990s due to increases in both student enrollment and teacher recruitment rates, recruiting, screening, and hiring practices are likely to have a much greater impact on a school district's ability to attract skilled science and mathematics teachers than is the case in the late 1980s. Districts that are able to offer attractive salaries and working conditions, to recruit aggressively, and to make offers in a timely fashion will be much more successful in attracting skilled teachers than districts that cannot. As a result, the variation in practices that currently exists may result in significant disparity in the quality of new hires attracted to different school districts. A particularly disturbing aspect of this prediction is that districts serving large numbers of disadvantaged children tend to have hiring practices that do not contribute to attracting skilled teachers. District practices through which seniority rules may restrict new hires to the least desirable school in the district or which introduce a long waiting period before vacancies can be opened to outside applicants are disincentives, as is uncertainty of initial school assignment. Consequently, increased competition for skilled teachers is likely to result in an additional factor contributing to the set of reasons that such children tend not to receive high-quality education.

A second implication of the qualitative evidence described in this section is that increasing teachers' salaries, although perhaps a necessary condition for attracting more skilled teachers to individual school districts, is not a sufficient condition. For example, it is unlikely that significant salary increases in the second district Shivers studied would lead to improved school faculties unless screening practices are reformed.

One might argue that these two implications drawn from the case studies are not particularly relevant to assessments of the adequacy of teacher supply and demand models. This would be correct if the goal of the models is seen in terms of assessing the overall balance between the demand for teachers and the supply of teachers. However, to the extent that the models are used to measure how well all school districts are able to provide qualified mathematics and science teachers to all children, then differences in the practices school districts use to recruit, screen, and hire teachers are extremely important.

SUMMARY

We have taken a close look at the effects of incentives to pursue a teaching career in science or mathematics on the supply of teachers. The factors influencing the individual's choices are beyond what the normal projection model can capture. Although better behavioral models are needed to measure the sensitivity of teacher supply to incentives, a number of research issues will have to be investigated before the models can be developed.

In addition, although the teacher supply and demand models considered in the panel's interim report do not use school districts as units of analysis, many decisions are made at the district level that affect supply and demand. As our case studies and interviews with school district personnel directors have shown, school districts vary greatly in the initiative they exert to fill their demand for teachers of subjects experiencing shortages. Individual maneuverability in recruiting and special or external circumstances affecting a district are key factors that influence a district's science and mathematics supply and demand situation—and these factors may outweigh those factors that can be quantified for modeling. These realities are central to the workings of supply and demand for science and mathematics teachers and should be monitored to the extent possible, as described in the following chapter and in Chapter 6. Chapter 6 concludes by recommending a series of conferences with a sample of school districts held on a regular basis, to discuss these factors and explore ways of recognizing them in statistical and descriptive reports on teacher supply and demand.

4
Monitoring the Supply Pool of Science and Mathematics Teachers

In this chapter we look at the models used at the national and state levels for projecting the supply of teachers for the next year, the components of these models, and what we would like to know to monitor more effectively the supply pool and its components.

The models examined in the panel's interim report (National Research Council, 1987c) were the projection model used by NCES and the teacher supply and demand models and projections developed in six states: California, Colorado, Illinois, New York, Florida, and South Carolina. They are most commonly projection models, which attempt to project teacher supply and demand and reach conclusions about surpluses or shortfalls of teachers at a point in the future.

Useful models should incorporate four major characteristics. The first is behavioral content, by which we mean models of relationships between variables in the environment and the behavior of actors in the education system. An example of a behavioral component in a model of teacher supply would be the estimated impact of salaries and working conditions on the decision of teachers to continue or to leave teaching. The models examined are limited by the lack of behavioral content. The second major characteristic of useful models is disaggregation by geographic area and subject field, and the third is quality measurement. Some of the models examined in our interim report incorporate useful refinements, such as the use of age-specific and field-specific attrition rates in projections of continuing teachers. But one key problem is a lack of useful geographic disaggregation. Moreover, only about half the state models examined disaggregated data by subject field). Nor do models deal in a satisfactory manner with the issue of quality. When models consider this dimension at all, the definition of a qualified teacher is equated with certification.

Finally, useful supply models should include all the sources of supply. Among the state models and the National Center for Education Statistics model that were the central focus of the panel's interim report, none provides what we regard as a detailed analysis of the contribution of the various components of potential teacher supply. Most of the models ignore supply sources other than newly certified teachers or some equivalent. The NCES model until recently limited projections of new entrants to new graduates of teacher training programs. Other definitions that are used in state models include students enrolled in the state's education programs and the number of newly certified persons. NCES and some state models have more recently broadened the components of the teacher supply pool.

Among these state models, the California PACE model (Cagampang et al., 1986) represents the most fully developed analysis on the supply side, with projections of the supply of new entrants from four sources: (1) new or recent graduates of California credentialing programs, (2) new credential holders from out of state, (3) teachers entering from the reserve pool of nonteaching credential holders, and (4) college graduates who pass the California Basic Educational Skills Test and obtain emergency credentials. Because of inadequate data sources and the lack of knowledge of the supply behavior of the various new entrant components, however, the PACE model relies largely on extrapolations of historical hiring patterns in the state, which are not the same as projections based on behavioral supply relationships

Overall, it is the panel's view that current models of teacher supply and demand have very limited usefulness for defining education policy and consist of little more than plausible extrapolations of relationships that are largely based on cohort survival techniques on both the demand and the supply sides. None of the models has any serious behavioral content—i.e., on the relationship between changes in circumstances (e.g. salary, working conditions, pension benefits, economic recession) and changes in the numbers and kinds of people interested in obtaining teaching positions or in the numbers and kinds of teachers demanded by school systems. Since much of the research needed to incorporate behavioral content in supply models has not been done, the panel considers monitoring supply to be the best course of action. By monitoring supply we mean gathering data relevant to teacher supply periodically and monitoring trends in the data. In the short run, efforts are needed to improve the consistency, scope, and quantity of data available for monitoring teacher supply. Concurrent with monitoring, research should be conducted to support behavioral models. As research findings on the relation between the incentives discussed in the preceding chapter and supply become available and the relevant data bases are developed, resources can be devoted to behavioral modeling. This chapter assesses the data that are now or could be collected to monitor the supply situation of science and mathematics teachers in this country.

MONITORING POINTS ALONG THE SUPPLY PIPELINE

The number of teachers employed in schools nationwide is augmented each year by new graduates from teacher training programs, newly certified teachers who enter teaching from other pathways (such as collaborative arrangements with industry), and previously certified teachers who are not teaching but have chosen to reenter the profession. The number of teachers employed is diminished by attrition due to retirement and other causes. Thus, monitoring supply requires keeping track of changes in the supply pool over its various stages. It requires data on certification, on incentives that motivate people to apply for or accept teaching positions, on new hires, and on attrition and retention rates.

NCES recognized the importance of statistics to monitor teacher supply and demand in the United States and initiated the Schools and Staffing Survey (SASS), an integrated set of questionnaires that are designed to provide several of the types of information sought. SASS is described in greater detail in Appendix B, along with descriptions of other national data sets. The first SASS questionnaires (for school districts, schools, school administrations, and teachers) were fielded in school year 1987-88. A follow-up survey, of all teachers in the base year who left teaching and a subsample of teachers who remained in teaching (both those who remained in the same school and those who moved to another school), was conducted in the 1988-89 school year. Thus, not only are current teachers included in SASS; there is follow-up information from subsets of teachers who left and teachers who remained. If SASS produces the data sought, the survey will provide the most valuable data related to teacher supply and demand the nation has had. As with all new surveys, some skepticism is in order about the ability of SASS to meet all its goals.

A useful way to envision components of supply that should be monitored is to identify stages along a pipeline, as outlined in Chapter 3. At the beginning stage are college students planning to teach. The pipeline progresses through degrees earned and certification, the decision to enter teaching, through retention and attrition rates.

One could add high school students' aspirations to become teachers at the very beginning of such a pipeline. However, although a high school student's expression of interest in a future career generally indicates the degree of regard for that kind of career or calling, it is probably not a reliable indicator of actual career choice. The Office of Technology Assessment's 1988 report, *Elementary and Secondary Education for Science and Engineering*, usefully describes a pipeline model that includes precollege students' views of and preparations for science and engineering careers (Office of Technology Assessment, 1988:6-20). It is used descriptively, however, and not for statistical modeling purposes. It states (p.6) that "students' intentions remain volatile until well past high school, with substantial numbers entering the pipeline (by choosing science and engineering majors) by their sophomore year of college." For purposes of monitoring data and generating information on the supply of science and mathematics teachers that could possibly be used in models, we begin the pipeline at the college level.

College Students Planning to Teach

The proportions of students enrolled in postsecondary education who are majoring in education; in mathematics; in computer science; in physical, biological, and earth sciences; and in engineering are key components of the supply of science and mathematics teachers at this early stage of the pipeline. The number of education majors is in decline, and many who may wish to teach now pursue a subject major. Therefore, monitoring the number of education majors provides only a partial count of this component of supply.

To gain a better understanding of the input to the pipeline, it would be desirable to make fuller use of the data that exist on freshman aspirations. Data from The American Freshman survey of the population of freshmen in higher education (described in Appendix B) include intended major and career aspirations and can be analyzed by sex and ethnicity as well. Data from the occasional follow-ups of those who have remained in college after two years and four years could be used to assess the value of freshman aspirations in judging changes in input to the pipeline. It would also be useful to have trend data on how many students who majored in the subjects noted above (subjects that are the most likely source of science and mathematics teachers) and who planned or did not plan to teach actually did or did not obtain certificates. The High School and Beyond longitudinal survey, which began in 1980, and the surveys of Recent College Graduates (both conducted by NCES and described in Appendix B) can provide these data for the science and mathematics majors in the samples of students who were surveyed.

Research conducted by individuals using these data sets can address issues in focused and informative ways. For example, Maxwell (1986) used data from The American Freshman and followed a sample of 2,000 freshmen to their junior year, relating their intended majors (education, not education), intended careers (teaching, not teaching), and declared major as juniors to their high school grade point average and rank and college grade point average. His findings (that the group who had not intended to major in education but planned on a career in education and later declared an education major had higher college grades than did other groups), although more related to quality, illustrate a wealth of information that could be used for research into supply questions, by subject major, at the postsecondary stages of a supply pipeline. In Chapter 6 the panel recommends measures to make these and other data more accessible to researchers.

Certification

Certification requirements vary from state to state, and the certification requirements in one state might not meet the requirements of another state. In some states a bachelor's degree that includes certain courses carries with it a teaching certificate. In other states, a year of teacher training beyond the bachelor's degree is required. In some states, as much as a master's degree (in teaching or in a subject discipline) is required for certification. In addition, 21 states now allow alternative certification routes (such as through a cooperative program with the military), and usually to staff particular subject areas with shortages (McKibbin, 1988). The meaning of national data on the number of certified teachers is somewhat ambiguous because of the diversity of states' certification requirements.

To quantify the pipeline leading to certification it is useful to look at the number of students enrolled in education programs. At present, the major source of such data is the American Association of Colleges for Teacher Education (AACTE), an association of approximately 1,200 member colleges and universities that have teacher education programs. Periodically the AACTE conducts surveys of small samples of member institutions to obtain data on the numbers of students enrolled in these programs.

During the panel's May 1988 meeting with personnel officers of seven large school systems, they suggested that it would be useful to school districts planning recruitment of teachers to know the number of people in the pipeline by field. State education agencies should collect data on the number of students and graduates preparing for certification to teach, by field and by type of program (i.e., traditional or alternative) and make these data available to districts on a timely basis. These data are indicators at one juncture of the pipeline of potential additions to teacher supply within the next year or two.

Changes in the actual number of graduates awarded certificates to teach science or mathematics should also be monitored. National data should be compiled from state certification board data on the number of new certificants—by type (regular, alternative, emergency) and by subject annually. As noted above, although different states have different certification requirements and classifications, it should be possible to estimate comparable totals for categories such as science and mathematics (elementary and secondary) and to present disaggregated data when available.

Comparability across states would be more achievable if proposals to standardize certification nationally through board certification (described in Chapter 5) are implemented. Since teachers may be certified in more than one category, these data will tend to overestimate the increase in the supply of newly qualified teachers. In fact, these data provide an upper bound of the change in newly qualified supply.

Information is also needed on the degree of reciprocity in certification across states. This information can be used to indicate the extent to which shortages in one part of the country could be filled by additional teachers from another. The effect of reciprocity is a research issue that relates to the mobility of the reserve pool. The National Association of State Directors of Teacher Education and Certification publishes states' reciprocity provisions periodically in its manual on certification. Reciprocity, too, could cease to be a problem if current proposals for national board certification of teachers come to pass.

Large proportions of new graduates of teacher certification programs do not go on to teach in the state in which they obtained certification. Why? Follow-up data on new certificants are desirable to ascertain the numbers and proportions of certificants who did not seek, were not offered, or did not accept teaching positions offered in their states. The follow-up questionnaire would probe for reasons why certificants did not teach, alternatives they pursued, and salaries. The survey of Recent College Graduates (RCG) is one possible source. It provides national data on graduates one year after receiving an education degree. The final survey in this series is scheduled for 1991, but it will be replaced during the next two years by a national longitudinal survey of college graduates. States may also find it valuable to follow their new certificants to understand loss to the pipeline of teachers at this juncture.

Data on the above aspects of certification would help to answer such questions as: How have increased state requirements for teacher certification affected enrollments in these programs or reentry after a gap in teaching? To what extent do states' restrictions and requirements placed on teachers moving from another state discourage them from reentering teaching in the new state?

Another occurrence that should be monitored as a possible indicator of shortage is the states' use of emergency or provisional teaching certificates, in science and mathematics in particular. According to the NEA survey Status of the American Public School Teacher 1985-86 (1987:20), only 8.4 percent of 1,291 respondents in all fields said they did hold such a certificate. (The data were not presented by subject area.)

Though the incidence seems minor, it would be desirable to have data from the states or local school districts, over time, on the number of teachers who hold temporary, provisional, or emergency certificates. SASS collects data from a sample of teachers on these types of certification in the teachers' primary, secondary, and best-qualified teaching fields. The design of the teacher sample will permit estimation of these rates of certification by type nationally and regionally and by type of district and school. Analyzed by subject, by region, or by type of area (e.g., rural, suburban, urban), these data may indicate exhaustion of the reserve pool or a shortage in a particular subject or in a particular geographic area. It would also be important to know how easy it is to convert a temporary certificate into a permanent one. Finally, it would be desirable to monitor the incidence of obtaining certification through various types of alternative certification programs. This could be done by adding a question on use of alternative certification routes to the questionnaire.

New Hires

Newly hired teachers come from many different sources, including new college graduates, former teachers, individuals who were certified but never taught, and teachers who change residence. School district personnel administrators indicated to the panel that typically a large percentage of their new hires were experienced teachers, not new certificants. Some administrators expressed a preference for experienced teachers. Descriptive statistics in many states indicate that a substantial fraction of new hires consists of teachers that fit into some category other than newly certified teachers. For example, less than 20 percent of new hires of mathematics and science teachers in New York State were new certificate holders; the corresponding figure in Illinois is 40 percent (National Research Council, 1987c:113,103).

An urban school district in a western state that took part in the panel's case study provides a more specific illustration of this phenomenon. There, only 4 of the 16 science and mathematics teachers hired in the past 5 years were new to teaching. And of those 4, only 1 came directly from college followed by a teacher training program; the other 3 had graduated between 3 and 10 years earlier and had recently gone back to school for their teaching certificate. Interviews with the teachers produced thumbnail sketches of previously experienced new hires:

- A newly hired mathematics teacher had taught science for 12 years in a junior high school in a neighboring suburban district and grew to dislike it. After other jobs for four years, he chose to return to teaching but preferred mathematics to science to avoid the responsibility and liability of labs.

- A newly hired chemistry teacher had taught for six years in a local parochial high school, following a career in the Navy where he did teaching and training full time. Without his retirement salary from the Navy, he would not be able to support his family on a teacher's salary. He was actively recruited by the district.
- A science teacher, hired five years ago, had been laid off three times, twice in this district and once in another district, all due to seniority. With six years' teaching experience, he was offered a position in five schools in this district five years ago.
- A mathematics teacher was hired on a part-time basis at her request. She had taught many years in another state, taught as a long-term substitute in this district, then as a regular teacher, and plans to return to full-time teaching next year after her children have adjusted to school.
- A teacher was newly hired in mathematics but in fact had taught chemistry at the same school for seven years. She began as a mathematics teacher, which was her major, but after teaching 12 years in this district was slated to be laid off. So she switched to chemistry (her minor) and waited for a math position to open.
- A mathematics teacher hired five years ago had taught at the junior high level in this district for eight years before moving to the high school level.

These examples and information from other states underscore the need to obtain data on all of the components that make up the new teacher supply.

Even new certificate holders may not be from the traditional new college graduate channel: they may be older people who left their occupations to earn alternative certification. A total of 21 states offers such a channel, mainly to meet the needs of particular shortage areas. In most of these states, alternative certificants constitute a small percentage of total new hires (McKibbin, 1988). But some states are notable exceptions (Carey et al., 1988:27-28): The New Jersey Provisional Teacher Program trained 240 mathematics and science teachers between 1985 and 1988. Among the districts that participated in the California Teacher Trainee Program in 1984-85 and 1985-86, the program accounted for 61 (15 percent) of new mathematics teachers, 101 (31 percent) of new biological science teachers, and 24 (24 percent) of all new physical science teachers.

To monitor the supply of science and mathematics teachers, it is necessary to be able to distinguish among the components of the corps of new hires. Ideally, to monitor supply at the district level, data are needed for the following categories by subject (chemistry, physics, biology, other sciences, calculus, other math, other subjects):

Certified teachers with no prior teaching experience:
- Certified before the last school year*
  - Out-of-state certificate
  - In-state certificate
- Certified during last school year
  - Out-of-state certificate
  - In-state certificate—whether through a regular or an alternative program

Certified experienced teachers:
- Teachers returning from temporary leave*
- Taught prior year
  - Out-of-state
  - In-state but out-of-district
- Did not teach prior year*
  - Last taught out-of-state
  - Last taught in-state but out-of-district
  - Last taught in-district

Noncertified college graduates with emergency credentials:
- Taught last year
  - In-district
  - Not in-district
- Did not teach last year*

*These categories came from the reserve pool.

Many of the districts in our case studies do not currently collect such data, and only a few (generally the larger districts) disaggregate the data by subject, as would be desired. SASS has been designed to collect such data.

Several other types of data could shed light on the supply of new hires. These include information on incentives to teach, reasons for selecting one's current school or district, the number of applicants per opening, the number of job offers per hire, and the extent of district recruiting. SASS presently asks school administrators how difficult it was in general (not for specific fields) to find qualified applicants to fill teaching vacancies last year. If the respondent notes it was difficult in some fields, space is given to write in those particular subjects.

Desired information regarding new hires includes the ability to monitor their incentives to teach. Comparative salary data are needed to indicate the competitiveness of beginning teachers' salaries relative to the starting salaries of alternative nonteaching positions. There is a question of just how this comparison should be made, but one simple measure would be starting salaries in industry for people with equivalent education, such as a bachelor of science degree in mathematics. The College Placement Council publishes these data annually (College Placement Council, 1988).

It would also be useful to know the reasons why science and mathematics teachers selected their current school or their current district and alternative offers they had. Such data are not currently collected by the SASS teacher questionnaire and were not part of the National Longitudinal Survey (NLS) teaching supplement. The data would help identify actions that schools or districts might take to attract well-qualified science and mathematics teachers. However, if there is a national shortage, such actions may only alter the geographic distribution of existing new teachers.

Trend data on the ratio of applicants to vacancies by field could be useful, even though it is recognized that the number of applicants for a position is a function not only of supply but also of aggressive recruiting and of the characteristics of the district and its schools. Although an applicant may apply for more than one vacancy, a decline in this indicator over time would point to increasing shortages of applicants in a particular field. Questions on the number of applicants could be added to SASS. Trend data on the ratio of job offers per hire could alert a district to changes in the attractiveness of its positions, but small (large) ratios could also reflect a surplus (shortage) of teachers.

Interviews with school district administrators revealed how shortages and surpluses in particular subjects were reflected in adjustments made in recruiting practices. A shortage in an area of need—frequently for minority teachers at all levels—often would trigger an aggressive recruiting effort, including trips to other states. It was often mentioned that a few years ago officials traveled to cities experiencing teacher layoffs to recruit science and mathematics teachers; today, however, the officials may have few or no vacancies in these subjects, and no recruiting is needed. It would be helpful to have data from school districts, perhaps building on SASS, on the extent to which districts are shifting from screening applicants to recruiting actively—by subject and by racial/ethnic group. When teachers are in surplus, districts recruit near home (if at all) and usually need only to screen and accept applicants. Thus, a count of the number of districts engaging in active recruiting is an indicator of shortage, and growth in these numbers over time suggests an increasing shortage.

The Reserve Pool

The reserve pool consists of people with teaching experience who did not teach last year, or individuals who were certified to teach at least a year ago but who have never taught. This reserve pool is a major source of new hires. The National Education Association estimates that more than half the new hires in the nation come from it (NEA, 1987f). In Connecticut, more than two-thirds of the new hires in 1984 came from the reserve pool. Returning experienced teachers constituted 55.5 percent of the new hires, and 12.0 percent were reserve pool members with no prior Connecticut public school teaching experience (Prowda and Grissmer, 1986:18). Since the reserve pool plays such a major role as a source of new hires, it is important to know how large it is and where it is found, or at least to know whether it is nearly exhausted. Concern about depletion of the reserve pool is not a state issue, but can be a special urban, rural, or regional issue related to the region's particular labor market situation. In major parts of the state of Louisiana, for example, it is impossible to find certified teachers to fill vacancies. Consequently, the schools are forced to hire on a temporary certificate or to drop courses from the curriculum. In these areas, not only is the reserve pool exhausted, but there are also insufficient newly certified teachers who are willing to teach under the conditions offered.

Little is known about who, among the various categories of people in the reserve pool, desires to enter or reenter the teaching profession. Thus, more important than estimating the size of the reserve pool is estimating the supply potential of the reserve pool, because some of the individuals in the reserve pool would not reenter teaching under any conditions. Different components of the pool can be expected to behave in very different ways. For example, teachers on maternity or health leave during a given year, or laid off and expecting to be called back, can plausibly be expected to return to the teaching pool in the next year at relatively high rates; newly certified teachers who did not obtain teaching jobs even though they have been in the market during the last few years, can be expected to remain in the teacher supply pool with relatively high probability; teachers whose credentials are older and who have been out of the teaching market for several years have a lower probability of being attracted back to teaching; while people with teaching certificates who have followed a completely different career path for many years have a much lower probability of being attracted to teaching. In some states, as we noted above, everyone with a bachelor's degree is potentially in the supply pool with some (arguably low) probability.

Several approaches could be followed to estimate the supply potential of the reserve pool: new college graduates could be followed over time, new hires from the reserve pool could be tracked backward, data could be collected and accumulated on the number of last year's certificants who did/did not get teaching jobs, and state agencies could use state certification files to study the reserve pool in the state.

Looking at these approaches in more detail, first consider what could be learned by following new graduates over time. The best source for this is data on the cohort of education majors in the Longitudinal Study of the High School Class of 1972. Heyns (1988) has used this data base to study entry and attrition of this cohort and to collect information on who left teaching and who wants to return (potential supply in the reserve pool). She found that, by 1986, one-quarter of the education majors who completed teacher training programs had never taught. Two other longitudinal studies, High School and Beyond and NELS:88, will provide similar opportunities for studying the reserve pool in the future. Another source of one-year longitudinal data is the Recent College Graduate (RCG) Surveys carried out by NCES periodically from 1976 through 1984. Using combined data from these surveys, the Office of Technology Assessment (OTA) analyzed the career paths of newly qualified teachers (1988:55). It found that at about one year after graduation 20 percent of newly qualified teachers had not applied for a teaching position, and that the remaining 80 percent who had applied for teaching jobs could be separated into 51 percent teaching full time, 11 percent teaching part time, and 18 percent not teaching at all. In fact, one year after certification, 49 percent of the certificants were in the reserve pool.

It is also possible to study new hires who come from the reserve pool, including both reentrants to teaching and individuals who have not taught but who were certified over a year ago. The SASS questionnaires for schools and teachers collect career information on both new entrants to teaching and experienced teachers regarding their recent career paths; these data will provide a rich resource for research on the reserve pool. It will be possible to study age of reentry, occupation just prior to reentry, the number of breaks in service, different patterns across different fields, and differences among different categories of members of the reserve pool. If the proportion of new entrants coming from different categories of the reserve pool (e.g., individuals certified in the past five years who are first-time teachers, returning teachers from the same district, returning teachers from the same state, returning teachers from out of state, etc.) are monitored over time, the trends will begin to shed light on the adequacy of the reserve pool to meet the demand. The SASS data are expected to provide an exciting opportunity for study of the reserve pool.

At the state level, an important segment of the reserve pool consists of individuals certified by that state who are not currently teaching. By the use of data from the retirement files and the certification files, it is possible for some states to track certificants who still reside in the state and to characterize that segment of the reserve pool by age, certification field, and years of teaching experience. A survey of these individuals can determine interest in teaching or incentives that would make teaching interesting to them. This would give some measure of the potential supply in this segment of the reserve pool.

A study of this type was carried out in 1987 for the Commonwealth of Massachusetts by the Massachusetts Institute for Social and Economic Research (MISER). MISER's *Report on the Status of Teacher Supply and Demand in Massachusetts* (1987) used certification files to determine the percent of those certified in Massachusetts in a given year who were not teaching the following year(s). Certificants were tracked for several years forward. It found that certificate holders were hired as long as 12 or 13 years after certification.

Connecticut maintains an ongoing activity to monitor its reserve pool of teachers. Part of that effort focuses on the nearly 3,000 new certificants who did not take teaching positions within a year following certification. (Nearly 4,000 had received certificates; it was found that only 992 of them were employed in public schools.) A random sample of 36.4 percent of these individuals was surveyed in 1987 (Prowda and Beaudin, 1988). Addresses were available for only 2,727 individuals certified in the state in 1985-86 but not teaching in 1986-87. Questions included current employment and salary, any prior teaching experience, undergraduate subject major and grade point average, teaching application experiences, and future employment plans. Certification files provided further information on subject specialties. Among the results: the median age of the nonteaching new certificants was 33; on average, they obtained their certificates five years after college graduation; slightly more than half had never taught full-time; the majority were employed, usually in education-related work (substitute teaching, enrolled as graduate students, tutoring, etc.); 52 percent had applied for public school teaching positions for 1986-87; only about 30 percent of respondents intended to apply for the 1987-88 year; and proportionately fewer science and mathematics majors would apply than education, social studies, and humanities majors. Salary sensitivity was analyzed using the questionnaire data: "The probability that a non-teaching new certificant holding a job outside of education applied for a public school teaching position declined by 6.6 percent for each $10,000 he/she earned in salary" (p.9).

The results of the Massachusetts and Connecticut studies are offered as examples of what can be learned about certificants who have never taught—an important segment of the reserve pool. While national data are not available on certificants who have never taught, states can be encouraged to gather such data, and NCES can compile and disseminate states' information accordingly.

Two research areas concerning the reserve pool warrant study. One is to study how much change in incentives would be needed to attract reserve pool members to teaching. The Connecticut study on salary incentives could be repeated in other states. In addition, studying the number of returnees to school districts in which there have been large salary increases (e.g., Rochester, New York) could reveal some information on the size of the potential supply in the reserve pool. Another topic would be the effect of limited mobility on entry to teaching. It appears that people will move a short distance but not a long distance to accept a teaching position. A study of the recruitment areas for urban, suburban, small town, and rural districts would help define the geographic areas in which members of the reserve pool represent potential supply.

In view of the major importance of the reserve pool to teacher supply and the concern that the other constituent of supply, newly certified teachers, is decreasing, the panel urges that all of these types of data collection and research activities be carried out.

Retention and Attrition Rates

New entrants to teaching are one major component of teacher supply. The other major component—and the largest—is the corps of continuing teachers. Continuing teachers represent typically 90 percent of teacher supply in any year. Models tend to use single retention rates or attrition rates for projection purposes. More successful state models use attrition rates that are differentiated by age or years of experience and by subject field. We reaffirm our support, as stated in the panel's interim report, for the use of timely, disaggregated data to determine the proportions of teachers who can be expected to stay or leave. Improvements can be built into SASS to ensure useful data on teacher retention and attrition by subject. We again acknowledge that there are tricky problems in using information on retention to project continuing teachers. For example, teachers who leave one school may simply transfer to another, and for a national portrayal of supply this kind of mobility needs to be subtracted out. Models for a subject such as biology need to be sure not to count as continuing biology teachers those who were teaching another subject last year.

In this vein, additional data are called for to monitor, over time, the important phases of the supply pipeline encompassing retention and attrition. To what extent do science and mathematics teachers leave teaching early in their careers? Do statistics indicate a large wave of retirements in the next five years? 15 years? It appears from an analysis by RAND of the teaching force in 1976-77, 1980-81, and 1983-84 that the proportion of teachers age 55 and over (9.5-10 percent) was quite stable. There was a clustering of secondary school teachers in mid-career, (age 35-44); 36.2 percent of secondary school teachers in 1983-87 were in this age group, an increase from 22.6 percent in 1976-77. The older members of this group are expected to become eligible for retirement in about 15 years (i.e., 1998-99). At that time the beginning of a wave of retirements may be anticipated (Haggstrom et al., 1988:8-9). For science and mathematics teachers, patterns of age, years of experience, and expected retirements seem to be similar. The 1985-86 survey of science and mathematics education by Weiss (1987) found that the typical high school science or mathematics teacher in 1985-1986 had 14 years of prior teaching experience, compared with 11 years for a typical teacher in 1977—some evidence of aging. The data do not support the prediction of an unusually large wave of retirees in the next decade. The class intervals in the Weiss study differ from those in the RAND study, although she found that only 14 percent of secondary school mathematics teachers and 16 percent of secondary school science teachers were over 50 years old (Weiss, 1987:64).

Data are thus needed from districts on the distribution of teachers by level, age, race/ethnicity, sex, and discipline. Attrition should be classified by retirement or other cause. Comparative salaries should be included to glean more about the competitiveness of teacher salaries relative to opportunity cost salaries. More specifically, information on the salary scale is needed, unless both average salary and years of experience are available. These data would enable the separation of attrition due to retirement from attrition due to incentives to leave teaching in favor of something else. The best prospect for obtaining these data is probably SASS, which included questions on attrition, by field, in the base year survey, but response to those questions was unacceptably small. The panel encourages efforts by NCES to modify the SASS matrix questions on attrition to simplify and improve response and to collect these data on a continuing basis. The school questionnaire should be able to separate attrition due to moving to another district from attrition due to leaving the teaching profession completely, which reduces the national supply of teachers.

Overall changes in supply are affected by factors that make teaching more or less attractive compared with other occupations. The SASS follow-up survey of former teachers, conducted in spring 1989 and to be surveyed again in 1991 and 1993, should provide data on this aspect on a national scale. Another rich source of data that are available for analysis is the supplemental questionnaire sent to over 1,000 past and present teachers and those trained for teaching in 1986 as part of the fifth follow-up of the National Longitudinal Study of 1972 (NLS-72) (see Appendix B for a description of this subsample). In a preliminary analysis of the detailed career histories of these current and former teachers, Heyns (1988) found that nearly half (44.7 percent) of those who had taught for at least a year were no longer teaching by 1986. She found attrition rates to be particularly high in the first three or four years of teaching, and often it was male high school teachers who left. Most of those who left were single and took another job directly after leaving teaching; that is, the primary pattern of nonretirement attrition was not women leaving for homemaking.

Another factor that can influence the retention rate for a school district or state is the portability of teachers' pensions. It would be useful to have comparative information from states on teacher retirement policies. Such information would be helpful for research relating retention rates, portability of teacher pensions and retirement policies, and particularly the sensitivity of attrition rates to the generosity of retirement provision.

The effects and the relative strengths of incentives such as retirement systems and those such as salary and professional development opportunities on attrition or retention are worth pursuing through better data gathering. SASS does ask school districts about the minimum age, years of service, and penalty associated with their retirement plans. Nonetheless, additional information is needed to improve the analysis of behavioral components at the retirement end of the supply pipeline.

Teacher Mobility and Interstate Migration

Little information is available on teacher mobility. Although teacher mobility in or out of a district affects supply at the district level, it affects supply at the state level only if the migration is interstate and affects national supply only if the teachers immigrate or emigrate. Effective monitoring of teacher mobility should start with separation data from the states, in the interest of avoiding double-counting. To be checked is whether the state can subtract "movers" from "leavers."

Little is known of the effects of interstate migration on the supply of teachers for a given state, but large systems the panel interviewed in Texas, Nevada, and Washington are noteworthy for major recruiting efforts out of state, as are two of the large-city districts represented at the panel's May 1988 conference, the Los Angeles Unified School District and the school district of Dade County, Florida. While some states may have information on the in-migration of teachers (see Table 4.1 for New York State), most states do not maintain such data; for those that do, the data may not be comparable with other states. The National Governors' Association (NGA) surveyed 15 states in 1987 for information on teacher mobility and teacher retirement system characteristics (NGA, 1988:9-10). Six states were able to provide information on the number of new hires who had taught in another state. The NGA concluded: "Clearly, this is another example of the need for better educational statistics. Nonetheless, . . . the number of experienced teachers moving into a state is large enough to be of concern but manageable enough for a retirement system to attempt to accommodate mobile teachers."

None of the states has data for out-migration to other states. One effort to overcome that deficiency is a project funded by the National Science Foundation through the Council of Chief State School Officers and the Regional Laboratory for Educational Improvement in the Northeast and Islands. Known as the Northeast Teacher Supply and Demand (NETSAD) study, it is a seven-state cooperative endeavor being undertaken by the Massachusetts Institute for Social and Economic Research. A primary

TABLE 4.1 New York State Public School Classroom Teachers: New Hires by Type and Location of Occupation in Prior Year, 1975, 1980, and 1985

Location/Occupation Prior Year	1985		1980		1975	
	Number	Percentage	Number	Percentage	Number	Percentage
In New York State						
Education	13,282	73.3	9,149	74.8	7,503	65.2
Student	1,749	9.6	986	8.1	1,972	17.1
Homemaker	1,267	7.0	1,246	10.2	1,226	10.6
Industry	458	2.5	173	1.4	167	1.5
Other	1,377	7.6	669	5.5	649	5.6
Total	18,133	100.0	12,223	100.0	11,517	100.0

Outside New York State						
Education	657	53.0	359	47.9	399	40.9
Student	261	21.1	186	24.8	355	36.4
Homemaker	86	6.9	66	8.8	60	6.2
Industry	61	4.9	28	3.7	22	2.3
Other	175	14.1	111	14.8	138	14.2
Total	1,240	100.0	750	100.0	974	100.0
Total						
Education	13,939	71.9	9,508	73.3	7,902	63.3
Student	2,010	10.4	1,172	9.0	2,327	18.6
Homemaker	1,353	7.0	1,312	10.1	1,286	10.3
Industry	519	2.7	201	1.6	189	1.5
Other	1,552	8.0	780	6.0	787	6.3
Total	19,373	100.0	12,973	100.0	12,491	100.0

Source: New York State Education Department (n.d.b).

concern of the project is to track the historical interstate migration of teachers within the region. It was borne of the interest of the Chief State School Officers in the seven states to establish a regional teaching certificate.

The SASS teacher questionnaire includes information regarding the occupation of teachers just prior to their current positions, which should be a resource for studying mobility at both the state and district levels. In addition, the SASS follow-up survey of teachers who leave the districts in the SASS sample, and further analysis of the NLS subsample of teachers and former teachers, should provide some useful data on the behavioral components of teacher mobility. Do they move for salary increases, for better working conditions, or because their family moves? A different perspective on the behavioral components of teacher mobility could be obtained from in-depth discussions with a group of district personnel officers. A series of in-depth conferences with school system officials is recommended in Chapter 6.

To summarize the discussion of this section, the supply of precollege science and mathematics teachers can be envisioned as a pipeline marked by a number of stages or decision points. These decision points should be monitored for a clearer understanding of the incentives underlying the decisions made by individuals as they move through college and the world of work. Key decision points that call for further information are college students' selection of majors and career goals; earning certification to teach and to become certified in certain subject areas; the decision either to apply immediately to teach or to pursue a nonteaching activity; knowing about the composition of the group of new hires; understanding the supply potential of teachers in the reserve pool; and knowing more about the decision to remain in teaching or to leave. At each stage we have noted data that are needed to provide a clearer picture of the supply of science and mathematics teachers.

A SPECIAL CASE: THE SUPPLY OF MINORITY TEACHERS

In the interviews conducted with school district personnel administrators across the country, they frequently mentioned a shortage of minority teachers. If concentrated efforts were made to increase the number of minority science and mathematics teachers in particular, an indirect result certainly would be an increase in the overall supply pool of teachers. In this vein, then, we summarize the available data on the particular problem of the supply of minority science and mathematics teachers.

Statistics collected by the National Education Association (NEA) indicate that the proportion of minority teachers fell significantly between 1971 and 1986. In 1971, 8.1 percent of the teaching force were black and 3.6

percent were of other minority groups. By 1986, however, black representation had dropped to 6.9 percent, and other minorities had declined to 3.4 percent (NEA, 1987e:14). Comparisons with minority school enrollment proportions show how seriously under-represented minority teachers are. A policy statement issued in September 1987 by the American Association of Colleges for Teacher Education (AACTE) summarized data from the NCES on student enrollments (CES, 1987b) and from the NEA on teachers (NEA, 1987e) in a clear illustration of the problem (AACTE, 1987a:3):

- Blacks represent 16.2 percent of the children in public schools, but only 6.9 percent of the teachers.
- Hispanics represent 9.1 percent of the children in public schools, but only 1.9 percent of the teachers.
- Whites represent 71.2 percent of the children in public schools, but 89.6 percent of the teachers.

To trace the special problem of minority underrepresentation in precollege teaching—and, when data are available, science and mathematics teaching in particular—selected statistics are shown below to illustrate the monitoring of minority-teacher supply at eight stages of a supply pipeline.

1. *Minority enrollment in higher education.* The data on minority enrollments that follow were compiled by NCES and reported in its 1988 report Trends in Minority Enrollment in Higher Education, Fall 1976-Fall 1986 (CES, 1988d). Black enrollments in higher education went from 1.03 million in 1976 to a high point of 1.11 million in 1980 and declined somewhat to 1.08 million as of 1986. Enrollment for other minority groups has increased steadily since 1976: Hispanics (from 400,000 to 625,000 in 1986) and Asians/Pacific Islanders (more than doubling from 200,000 in 1976 to 448,000 in 1986). Black male enrollment has undergone the most significant rate of decline among minorities, having fallen about 7 percent between 1976 and 1986. In terms of proportions, minorities generally constituted approximately 17.9 percent of total enrollment in institutions of higher education as of 1986—an increase from 15.4 percent of total enrollments in 1976. For blacks, however, the proportion had fallen from 9.4 percent in 1976 to 8.6 percent in 1986.

2. *Interest in majoring in education.* For college-bound black high school seniors who noted their intended college major on their SAT background information forms, interest in majoring in education decreased between 1981 and 1984. Baratz (1986), citing the yearly profiles of college-bound seniors published by the College Board, reported that in 1981, 5 percent of black student respondents intended to major in education. But by 1984 only 3.4 percent of black students intended to do so (pp. 9-10). And when the mean SAT scores of black high school seniors by intended major are compared, the highest mean SAT scores are found among students

headed for the fields of engineering, physical science or mathematics, and biological sciences. The mean SAT score for students who were planning to major in education was the lowest among the eight fields of study (Baratz, 1986:8).

3. *Bachelor's degrees earned.* It appears that the minority teacher supply pool continues to decline with each stage along the pipeline. Most of the declines occur among blacks; other minority groups seem to continue in the supply pipeline at greater rates than blacks, and at greater rates than a decade ago. As reported by NCES in Education Indicators—1988 (NCES, 1988f:104):

> Blacks earned fewer degrees in 1985 than in 1977 at all degree levels except the first-professional (e.g., M.D., J.D.). The declines are particularly significant when compared with increases in the young adult black population during the same period: it rose 7 percent among 18- to 24-year-olds and 40 percent among 25- to 34-year-olds. Men accounted for nearly two-thirds of the drop in degrees [among blacks] . . . . Hispanics, Asians and American Indians/Alaskan Natives earned more degrees in 1985 than in 1977 at all levels. The increase among Hispanics was in line with their population growth.

The number of bachelor's degrees in education (though this statistic has limited meaning given the movement toward requiring subject majors) fell significantly overall, from 143,462 in 1976-77 to 87,788 in 1984-85—a 39 percent decline. But for blacks it declined particularly steeply. In 1976-77, 12,922 blacks earned bachelor's degrees in education; in 1985, only 5,456 did—a 58 percent decline (NCES, 1988f:290,292).

4. *Pursuing a master's degree in education.* In an NCES special report, Hill (1983:18) reported a large decrease between 1976 and 1981 in the number of education degrees earned at the master's and doctor's degree level as well as the bachelor's degree level. Fewer students as a whole obtained an advanced degree in education, but for blacks the number declined more than for graduates in general. Overall, 128,417 graduate students received master's degrees in education in 1976 (NCES, 1986:130), by far the most popular advanced degree pursued (next was the master's in business, with 42,512 recipients). By 1985, only 76,137 master's degrees in education were earned—a dramatic 41 percent decline (NCES, 1988b:211). For blacks, however, there was a 53 percent drop in the number receiving master's degrees in education: from 12,434 in 1975-76 (Hill, 1983:27) to 5,812 in 1984-85 (NCES, 1988b:223).

5. *Pursuing teaching versus other endeavors.* While the number of blacks earning bachelor's and master's degrees in education fell substantially between the mid-1970s and the mid-1980s, the total number of

first-professional degrees and M.B.A.s awarded to blacks rose during this period (Hill, 1983:27,29; NCES,1988b:223,228). In a *Phi Delta Kappan* article documenting the shortage of black teachers, Patricia A. Graham cited a variety of data sources indicating decline in the supply of minority teachers. Graham speculated about the alternative routes black students in higher education could be taking (Graham, 1987:603):

> Experts differ about the causes for the decline in the number of black teachers and for the decline in the number of black students seeking to major in education. The first explanation, favored by those who believe that America is slowly but inevitably progressing toward racial justice, stresses the broader range of career choices available to educated blacks today. Previously, teaching was one of the few jobs available to college-educated black men and women. Because we now see blacks in other professions, we conclude that these successful middle-class blacks would have been teachers in previous generations. To some extent, this conclusion is probably correct, but not enough blacks have moved into other professions as yet for us to be certain that they represent displaced teachers . . . . Young blacks who are choosing alternatives to careers in education are not shifting in significant numbers to other professional fields.

The data we have available do not probe the alternative careers that minorities are pursuing, and we believe this will be an important research topic to study. The Survey of Recent College Graduates asks for the respondent's race. This survey may hint at decisions made immediately after college graduation, but it does not probe as deeply into alternative career or schooling decisions as would be required to analyze fully this special case of minority science and mathematics teacher supply. Data from the follow-up surveys of the NLS-72 sample could be analyzed by race/ethnicity to identify the career patterns minority students are selecting and the career patterns they pursue.

6. *New hires.* The number of new hires into the nation's public school systems is not known nationally by racial/ethnic group, but it should be available from SASS since the teacher questionnaire asks for the respondent's race and subject taught. SASS is also expected to provide data on what the person had done before entering teaching (though it does not ask for the salary of the job held prior to entering teaching).

7. *The current teaching force.* Among the general public-school teaching force, the percentage of teachers who are black decreased from 8.1 percent in 1971 to 6.9 percent in 1986 (NCES, 1988b:70). The proportions of secondary science and mathematics teachers in 1985-86 who are black are lower still. Among grade 10-12 mathematics teachers, only 3 percent are black; among grade 10-12 science teachers, 5 percent are black (Weiss,

1987:63). Most of the black and Hispanic teachers of science and mathematics are in the elementary grades. The Office of Technology Assessment (1988:58) posits: "For now, the proportion of minorities in the teaching force is increasing slightly, but several commentators warn of future shortages of minority teachers, particularly in mathematics and science."

8. *Retention and attrition.* Of 308 minority teachers surveyed as part of Metropolitan Life's 1988 Survey of the American Teacher, 40 percent said they were likely to leave teaching within five years, as opposed to 25 percent of the 891 nonminority respondents (Metropolitan Life, 1988:22). This may reflect the fact that 29 percent of the minority teachers surveyed for the Metropolitan Life project worked in inner-city schools, as opposed to only 9 percent of the nonminority teachers. The Metropolitan Life report summarized its findings on this aspect of teaching as follows (p. 5):

- Almost three out of four of the dissatisfied minority teachers say they are likely to leave, compared to about half of the dissatisfied non-minority teachers.
- Even among minorities who are very satisfied with their careers as teachers, more than one out of five say that they are likely to leave.
- Less experienced minority teachers are the most likely to say that they will leave. Fully 55 percent of minority teachers with less than five years of teaching experience say that they are likely to leave the profession.

Little information is currently available on retention and attrition rates of minority teachers. A study by Kemple (1989) of the career paths of 2,535 black teachers in North Carolina finds that, on one hand, over two-thirds of these teachers, who began their teaching careers between 1974 and 1982, stayed in teaching through 1985-86, and over one-third who had quit returned. On the other hand, the likelihood of teachers in this group leaving has been increasing since the mid-1970s, suggesting that as blacks gain access to other professions they may shorten their teaching careers. Kemple stressed that "even minor increases in attrition will have large influences on the overall representation of Black men and women in the teaching force" (p. 2).

Data from the 1989 SASS follow-up questionnaires of teachers who left and teachers who remain, which are expected to be available in 1990, and analysis of the 1986 survey of teachers and former teachers who took part in NLS-72, offer the best possibilities at this time of national data capturing much-needed information on alternative jobs taken and opportunity costs associated with teaching science and mathematics. These questionnaires ask for the respondent's race, subjects taught, length of experience, comparative salaries, and decisions made regarding continuing or leaving teaching, the pursuit of more schooling, or taking other positions.

Research is needed on the employment patterns of minorities and especially the difference in response to opportunity by minority women. Would increasing salaries attract them to teaching? Or is the cost of teacher training the problem? From 1958 to 1965 state teacher colleges provided blacks with easy access to education. They could live at home while going to college and could get a degree with the opportunity for income. A researchable question is whether the cost of teacher training relative to other occupational training has increased.

In conducting such research, analysis of existing data by racial/ethnic characteristics would be a productive first step. As a companion to the description in Appendix B of national data sets relevant to teacher supply and demand, NCES has published a useful compilation identifying six minority student issues and relating them to 32 NCES surveys containing racial/ethnic data (NCES, 1989b).

In conclusion, the troubling evidence thus far on minority science and mathematics teachers suggests a disproportionately acute shortage of blacks. The most valuable activity we can suggest—from analysis of data from SASS and other data sets and through the conduct of further research—would be to probe into alternative decisions black college students and graduates make, alternative positions they pursue, and the opportunity costs they perceive that draw them away from teaching science or mathematics.

SUMMARY

We have called attention to important gaps in the content of current state and national models. To understand fully the forces, influences, and incentives that affect the supply of science and mathematics teachers requires data and behavioral content that the current models do not capture. A sequential approach toward the goal of improved national models is thus recommended. For the short term, efforts can be made to monitor the state of supply—by further analyzing existing data such as those from the NLS-72, by building on the promising work of SASS, and by compiling and disseminating states' data, for a clearer portrayal of the supply situation in this country. This chapter described existing and proposed data that can serve as components of an effective monitoring system.

5
Statistics Related to the Quality of Science and Mathematics Teaching

As noted in previous chapters, the supply and demand for teachers of mathematics is brought into equilibrium in the short term by adaptations in the selection criteria for teacher or teaching quality. Thus a school system unable to hire science and mathematics teachers at a preferred quality level will have to lower its minimum quality requirements. Conversely, school systems facing a supply of teachers of acceptable quality in excess of the number they need will be able to choose those at the top of their quality scale, thus ending up hiring teachers of higher quality than suggested by their minimum criteria. While this comprises a generally accurate description of school system hiring practices, it does not tell us anything at all about what factors go into quality teachers or quality teaching. It is to that topic that we now turn.

It should be recognized from the beginning that we do not have very precise notions about what constitutes teacher or teaching quality, and thus we cannot provide definitive prescriptions as to types of data that need to be obtained in order to monitor either the level of teacher quality that exists or changes over time in quality. The problem is that assessment of quality is an extraordinarily difficult enterprise, and existing research does not go very far in identifying the factors that determine quality. It is the panel's view that the right dimensions of teacher or teaching quality are factors that produce a positive influence on student outcomes—that is, higher quality in our view should be defined to mean better student outcomes, given the influence of other forces besides teachers or school system factors that influence student outcomes.

Perhaps the best way to summarize the current state of knowledge on this topic is to note two sets of facts that come from existing studies of teacher quality.

1. Teacher quality matters a good deal to student outcomes, in the sense that it is possible to identify teachers who have produced well below average outcomes. In this context, *identify* simply means that specific teachers can be shown to produce relatively good outcomes, and other specific teachers can be shown to produce relatively poor outcomes (Contra and Potter, 1980).

2. If one tries to describe what factors are associated with teachers who produce good outcomes or bad outcomes, one finds very little association between particular characteristics of teachers and the resulting student outcomes. That is, better formal credentials, better preparation in terms of course work, more years of teaching experience, better scores on standard tests of teacher qualifications, etc., do not generally show up as teacher characteristics that are strongly related to better or worse outcomes (Druva and Anderson, 1983; Hanushek, 1986, 1989). It has often been found that teacher verbal ability is positively related to better student outcomes, but the relationship is not exceptionally strong; most other factors do not show up at all (Darling-Hammond and Hudson, 1986).

In sum, we know that there must be characteristics of teachers or of classroom situations that produce better student outcomes, and qualities or characteristics that produce worse student outcomes, but we do not know what these characteristics or qualities are with any degree of assurance.

Although it may be surprising to some readers that so little is known about what factors are related to teacher or teaching quality, a little reflection suggests that it is not so unusual that the state of knowledge is so limited. If one were to ask whether some people are more effective social workers and others less effective, whether some people turn out to be very successful business executives and others less so, or whether some people are very successful at doing survey research interviews and others are less successful, the answer in all these cases will surely be that there are very large differences in the degree to which people are successful or unsuccessful in particular kinds of professional activities. If one goes further to ask what factors are associated with success in being a social worker, a business executive, or a survey research interviewer, the answer will commonly be that very little is known about why some people succeed and others fail. The probable reasons are that the factors making for success are complicated, that personal characteristics and characteristics of the particular environment interact and may be idiosyncratic to particular situations or types of work environments, and that success has a lot to do with motivation, energy, striving for success, interpersonal skills, and myriad other factors that come together in subtle ways to produce better or worse outcomes.[1]

Given this state of knowledge, what should be done about the collection of data that relate to teacher or teaching quality? It is the panel's view that, although little is known about what factors are importantly related to quality, something is known about the kinds of factors that probably play some role in determining quality. We should try to collect the best such set of factors, recognizing that the data collected will not be sufficient to do a satisfactory job of explaining student outcomes. Thus in this section we discuss a number of types of data that are probably related to quality, although they have not been convincingly shown to be either strongly or systematically reliable indicators of quality. These results may be caused by systematic errors: for example, the better teachers teach higher-order skills, but tests measure primarily lower-order skills, so the quality difference in teaching is not measured.

The reader will note that we have talked about quality both in terms of *teacher* quality and *teaching* quality. The two are not synonymous. By teacher quality we mean those personal characteristics of individuals that enable them to be more effective in classroom settings: education level, subject matter knowledge, interpersonal skills in working with students, degree of inservice training, formal credentials, etc. By teaching quality we have in mind a somewhat broader notion that encompasses not only teacher characteristics but also the school setting in which classroom teaching takes place. Thus teaching quality includes factors that are beyond the control of the individual teacher: disciplinary norms of the school system or of the building principal, support given by principals to teachers, the presence or absence of inservice training opportunities or opportunities for interaction among teachers, types of textbooks that are selected for use in the school systems, amount of time allocated to each subject, number of classroom hours taught, and so on. Thus, teaching quality encompasses factors that

[1] The nature of the problem is illustrated by the example of survey research interviewing. This subject has been studied for many decades, and what we know with certainty are only a few relevant facts, none of which is sufficient to design a test to predict success at survey research interviewing. There are enormous differences in degree of success. Some interviewers achieve close to a 100 percent cooperation rate and have virtually no refusals, collect consistently high-quality data, and do so with relatively few hours expended in the interviewing task and thus have lower costs. Other interviewers have extremely high refusal rates, do not collect consistently high-quality data, and take a great many hours to produce relatively mediocre results. Although we know that these differences exist, it has not been possible to identify personal characteristics that would enable survey research organizations to predict who will be a good interviewer and who will not. Conventional demographic characteristics (educational level, experience, age, etc.) are of virtually no use in explaining success. Although a few personality characteristics seem to have some association with success, the state of knowledge is still relatively crude, despite a great deal of methodological work.

are not within the control of individual teachers, while teacher quality includes only those factors that relate to the personal characteristics of individual teachers.

In examining the quality of mathematics and science teachers, we have in mind a broader notion than assessing the quality of teachers who specialize in mathematics or science. Although some districts employ teachers who specialize in science or mathematics as early as the fourth grade, most teaching in mathematics in grades K-8 is done by teachers in either elementary or middle school who may not be classified as science or mathematics teachers, but rather as teachers who teach science and mathematics. The distinction is important: we are interested in assessing the quality of mathematics and science teaching on the part of teachers who teach those subjects, and many of them—probably most—are not specialized in the teaching of either science or mathematics.

Moreover, we are also interested in those dimensions of quality that relate to preferences of the school systems for the types of teachers they wish to hire. It is clear enough from our case studies, as well as from extensive discussions with the personnel directors of large city school systems, that mathematics or science teachers are not hired solely for the perceived quality of their mathematics or science teaching. Many school systems have other dimensions of teacher performance in mind when they hire teachers. In some school systems, the ability to fit in with the community is important; in some, the ability to teach other subjects or to direct extracurricular activities is important; in some, the ability to work with the types of students in the school system is perceived to be extremely important. The basic point is simple enough: school systems do not hire teachers to teach science and mathematics solely because of their perceived ability to be effective in classroom settings. Rather, hiring decisions are influenced by a great many other factors, some of which will necessarily result in hiring people who are likely to be less effective in teaching science and mathematics than teachers who were not hired because they lacked other skills or characteristics.

In the remainder of this chapter, we attempt to sort out the major ingredients of teaching and teacher quality that call for further data. We look first at school system policies and practices and the school-level conditions that can affect teaching quality. Next we look at the qualifications of incoming teachers—their college and professional preparation, their level of achievement in science and mathematics, their cognitive abilities, and so on. Finally, we examine other factors that also influence student outcomes but do not fall neatly under either school system policies and practices or teacher qualifications and characteristics: curriculum and textbook selection issues, time-on-task issues, and issues relating to the home environments of students. All of these do or may influence student outcomes to a substantial degree, and none is likely to be under the control of either the teacher or the school principal.

SCHOOL SYSTEM POLICIES AND PRACTICES

The assignment of a teacher to courses and pupils appropriate to the individual's educational background, certification status, and experience is crucial to quality instruction in precollege mathematics and science. But district personnel policies, budget constraints, and other external factors can impede the ability to achieve the most effective match.

A policy maker with the specific goal of higher-quality instruction often finds that it is difficult to change many of the policy variables that affect the quality of instruction. District policies exist in a complex web of competing goals and pressures. Even if the central goal is quality teaching, the policy maker must also consider school system policies and union contract provisions regarding recruitment, initial assignment, and transfer and retention of teachers. A given set of policy guidelines can have quite different effects depending on whether enrollment is stable, growing, or declining. For example, seniority rules for assignment or transfer have different effects in environments in which enrollments are rising or declining. Personnel policies are also affected by the enrollment size of a particular school system, the enrollment size of a high school, the extent to which the curriculum is taught by specialists, and the match among educational background, teaching assignment, and teacher and student cultures.

Recruitment and Hiring Practices

Certain policies set by the school district, teacher organization, or state school finance plan can have deleterious effects on the ability to hire the most talented teachers. The examples given here apply not only to science and mathematics teachers but probably also to teachers in general.

Discussions with personnel offices of large school systems suggested that recruitment of new teachers by large districts with diverse student populations was often hindered by the fact that recruiters could not specify the school to which the applicant would be assigned. Many persons would find such a school system desirable only if they could teach in a given section of the school system or in a specified school. Since recruiters could not make such commitments, or could not make those commitments early enough in the recruitment period, candidates were lost to the school system. This problem stemmed from district policies related to the timing of hiring, interviewing, and specific placement. District policy in some systems requires the applicant to be interviewed only by the district administrator; subsequent assignment is a central office decision. Other district administrators screen applications and refer promising candidates directly to principals, who conduct the interviews.

The uncertainty of initial assignment also seemed to be exacerbated by seniority rules of internal transfer. In one medium-sized school district in a western state that participated in our case study analysis, internal transfer rules took months to implement. With a tendency for junior high science and mathematics teachers to request high school positions, and for elementary teachers to request junior high positions, the process of considering all transfer applications and then determining which positions were actually vacant continued well into the summer. Job offers could not be made until August. Since other districts could make job offers in March and April, this district was left with candidates who had not obtained positions elsewhere.

In some circumstances, the problems stemming from seniority rules become especially severe when combined with rehiring rights after teachers have been laid off due to enrollment decline or financial constraints. In such circumstances, district rules, regulations, and practices rather than professional judgment often seemed to determine the match between teacher and classroom assignment. For example, seniority rules may restrict new hires to the least desirable schools in the district. These rules may drive teachers not only from the school system but also from the profession. Seniority rights may also prevail when teachers are transferred among schools. When vacancies occur, the teacher with the greatest longevity in the school system may have first choice. When enrollment declines, teachers with higher longevity in the school system, the school, or a teaching field may have rights to bump less senior teachers. The length of the waiting period before opening vacancies to outside applicants greatly affects the district's ability to sign on talented applicants. Many officials said they lose good applicants to other districts whose rules or budgets allowed them to hire sooner.

Enrollment size and composition also influence district policies. The hiring restrictions of one large urban school system in the West contrasted starkly with the innovative practices for meeting future needs employed by a small suburban school system in the same region. The suburban superintendent, in conjunction with a nearby college, recruited well-trained graduates to fill projected vacancies. The smaller enrollment size and relative wealth of the suburban school system, as well as the homogeneity of the student population, accounted for the differences in practices between the suburban and the urban systems. In another suburban school system in the East, a teacher who attracted high school students to advanced science classes had been allowed to develop his own teaching assignment. Such flexibility is less likely in a larger school system concerned with uniform course offerings among schools. One of the reasons for more rules, and sometimes less flexible ones, in larger school systems is the need to adhere to goals of equity among staff members in conditions of employment.

Factors external to the school district can also affect local hiring practices. Increases in state-mandated graduation requirements for mathematics or science can cause the district to fill vacancies in those fields with teachers not yet certified in the particular subjects, in order to meet the state requirement.

As noted in Chapter 2, 42 states have added requirements in science or mathematics since 1983. The Center for Policy Research in Education (CPRE), which has surveyed the states' graduation requirements, has found that in schools affected, about 27 percent of students are taking an extra mathematics course and 34 percent an extra science course (CPRE 1989:33). Many of these students are middle- to low-achieving, the CPRE study relates (p. 35). CPRE inquired as to the nature or level of the additional courses. In many instances the added courses were remedial or lowerlevel science and mathematics courses (p. 35-36). The increased requirements undoubtedly have changed schools' staffing patterns and course assignments and have probably affected hiring practices for science and mathematics teachers.

State-mandated minimum competency test scores and state school-finance formula constraints on local funds for laboratory equipment and supplies, computers, teacher aides, or teacher salaries are other external factors that local personnel officials must take into consideration in hiring teachers. An unintended consequence of decisions made under these conditions may be a loss in teacher or teaching quality.

Of course, not all rules act to restrict supply or make the task of matching persons and assignments more difficult; certain rules may benefit some school systems. When there is a potential for future growth in high school enrollments, teachers in a school system may pursue advanced study so that they can move from elementary school or junior high to high school. Other teachers may be attracted to begin their career in the district with a thought toward future advancement. Without seniority rules, there would be no such encouragement, as new hires might occupy newly created positions in high schools.

Data are needed to better describe the incidence of these and other policies and practices that affect the ability to hire and place the most promising candidates to assure instruction of high quality. The Schools and Staffing Survey (SASS) does not yet provide data related to most of these areas. In-depth conferences with a sample of SASS districts on a regular basis are recommended (see Chapter 6) to gain more accurate insights into the use of such policies and practices.

Misassignment of Teachers

Teacher assignment is critical to quality instruction in all subjects, especially so for science and mathematics. Misassignment of science teachers can occur when a vacancy in a science specialty is filled with a certified science teacher who is unfamiliar with that particular field. High schools may be too small to have a full-time chemistry or physics teacher or even a full-time biology teacher.[2] In 1986-87, only 13 percent of teachers who taught physics in secondary schools had teaching assignments in physics alone. Almost two-thirds of the teachers who taught physics had their primary concentration of classes in chemistry, mathematics, or general and physical science (American Institute of Physics, 1988:17). There may be a need for one but not two science teachers. The same type of misassignment can occur in mathematics, when a teacher is trained to teach areas of mathematics other than that assigned or some other subject altogether. In many states, it is legal to assign a teacher to teach part time in an area in which the teacher is not certified, under a practice called out-of-field teaching as opposed to "misassignment" (Robinson, 1985).

Estimates of the prevalence of misassignment based on data from the early 1980s collected by the National Center for Education Statistics (NCES) and the National Education Association (NEA) vary considerably. In a preliminary report on indicators of precollege education in science and mathematics, the National Research Council (NRC) notes the erosion of the quality of the existing teaching pool by misassignment of newly certified teachers. This report cites NCES findings that, among bachelor's degree recipients in 1979-80 who were teaching elementary or secondary

[2] In one of the case studies, the employment of a full-time chemistry teacher by a school system was mentioned. This condition was treated as rare for the school systems studied. Such employment can be seen as unusual for the United States by examining some necessary conditions. If one assumes that a teacher teaches 5 classes and that a class has between 25 and 30 students, then to teach a single subject at the same grade level requires 125 to 150 students per grade level. For a 4-year high school this means a school enrollment size of 500 to 600. For a 3-year high school, it means an enrollment size of 375 to 400. In 1982-83 9.5 percent of secondary students attended schools below the latter size criterion. An additional 10.5 percent of secondary students met the former criterion. If only half of the students take a chemistry course, then slightly more than half of the students, 53.3 percent, attend such secondary schools. If only a third of the students take a chemistry course, then only slightly more than 10 percent of secondary students (13.4 percent) attend schools of that enrollment size (NCES, 1986:68). That only a third of secondary students are likely to take a chemistry course can be garnered from the fact that 65.4 percent of public secondary school students take natural science (p. 41), and the average number of Carnegie units (a standard of measurement that represents one credit for the completion of a one-year course) in natural science is 1.9 (p. 44). Expanding the ranges of possible courses in natural science to include two courses in chemistry or chemistry and physics would indicate that only 3.9 percent of schools, that is, the schools with larger enrollments that enroll 13.4 percent of the students, would be able to hire a full-time chemistry or physics teacher.

school full time in May 1981, only 45 percent of science teachers and 42 percent of mathematics teachers were certified or eligible to be certified in the field in which they were teaching (NRC, 1985:52). More recently, Darling-Hammond and Hudson (1987a:21) reported "estimates that vary depending on who is asked to estimate the degree of misassignment (school administrators versus teachers) and on how misassignment is defined." They reported (1987a:21):

- Not certified in area of primary assignment: 9-11 percent by teacher report, 3.4 percent from central office administrators' estimates (NEA, 1982; NCES, 1985a).
- Not certified for some classes taught: 16 percent by teacher report (NEA, 1982).
- Less than a college minor in area of primary assignment: 17 percent by secondary school teacher report (Carroll, 1985).

The 1985 National Survey of Science and Mathematics Education found higher proportions for science and mathematics—18 percent of grade 7-9 mathematics teachers and 14 percent of grade 10-12 mathematics teachers teach courses for which they are uncertified. For science teachers, the percentages are 25 for grades 7-9 and 20 for grades 10-12 (Weiss, 1987:77-88).

Transfer policies can sometimes lead to a misassignment and thwart a teacher's potential for advancement. In our contacts with school district administrators, a tendency was reported for principals to transfer teachers from subject fields of surplus to subject fields of need. Often, these transfers moved the teachers from their primary subject fields to different areas. Transfers of this nature took place due to changes in student demand for subjects under stable enrollments as well as in times of changing enrollments. Such transfers also occurred because principals sought teachers able or willing to handle extracurricular tasks such as athletics, the school paper, the yearbook, or student clubs.

The extent to which misassignment occurs today in science and mathematics may be greater than for other subjects. Data on the extent of misassignment for all fields at the school district level will be obtainable from the SASS Teacher Demand and Shortage questionnaire. It will also be possible to estimate misassignment by field by using the SASS teacher questionnaire. This questionnaire obtains courses currently taught by each departmental teacher, the teacher's area(s) of certification, and college major and minor. Estimates of misassignment by field as defined by certification status can be made using these data.

Since certification standards vary so much across states, the fact that one was not certified in the field in which one is teaching does not necessarily mean misassignment. To obtain a more complete picture of misassignment, information on inservice training and actual course-taking preparation should also be analyzed, as Darling-Hammond and Hudson suggest (1987a: 21-22). The SASS teacher questionnaire represents a promising step forward. It requests data not only on certification status (as above), but also on degrees earned and major and minor fields of study, amount of course work in primary and secondary teaching assignment fields, and, for teachers who teach any science or mathematics courses, the number of graduate and undergraduate courses taken in various categories. These are rich data to examine misassignment and out-of-field teaching.

Information from SASS should be analyzed together with state certification data on the number of emergency certificates issued in science and mathematics; 46 states allow emergency certification. Of these, 30 require university course work in order to renew and work toward full certification (McKibbin, 1988:32). Supplementary data would include state rules on the extent to which out-of-field assignment is legal. Such information from various sources, when analyzed jointly, will help monitor the extent and trends of misassignment in science and mathematics teaching.

Providing for Inservice and Continuing Education

Some of the most important district and school practices that affect the quality of instruction are those directed to teachers already in place. To maintain quality instruction throughout their careers, teachers require professional support from their schools and districts. This support includes working conditions, facilities such as laboratories, materials and supplies, collegial and administrative support, resources for continuing education, and opportunities to influence decision making (Darling-Hammond and Hudson, 1987a:27-37).

District practices regarding inservice and continuing higher education for teachers in place affect teacher quality directly and can make it more or less attractive for a teacher to continue in a district. School districts have been the primary sponsors of inservice programs, but such programs are highly vulnerable to district budget cuts.

Decisions as to what kinds of inservice education to fund with a limited budget affect teaching quality in ways that data alone may be unable to illustrate. In one large, suburban, low-wealth district we studied, much of the staff development budget was geared to weaker teachers. Teachers had little release time during the school year—17 days allotted for each high school. Only about 20 percent of staff development was used for college-level course work.

A national commitment to teachers' continuing education appears to be missing. The federal government does support inservice education through the Title II program of the Education for Economic Security

Act of the Department of Education and through the National Science Foundation (NSF) Teacher Enhancement Program (Office of Technology Assessment, 1988:69), but funding for both activities is severely limited. Appropriations for the Title II program have been uneven, dropping from $100 million in fiscal year 1985 to $42 million in 1986, then $80 million, $120 million, and $127 million in 1987, 1988, and 1989, respectively (OTA, 1988:123; U.S. Department of Education, 1989). These are small amounts when viewed on a per-pupil or per-teacher basis. The Office of Technology Assessment notes by comparison that a $40 million education program equates to a spending of $1 per pupil or $20 per teacher (1988:123). NSF's Teacher Enhancement Program funds a small program of teacher institutes emphasizing teaching techniques in science and mathematics. The institute program is much smaller than it was in the past. Between 1954 and 1974 NSF spent over $500 million on teacher training institutes that at their peak involved 40,000 teachers (OTA, 1988:119-120). The Teacher Enhancement Program has been revived somewhat since 1982, when it was virtually nonexistent. According to Charles Hudnall of the NSF staff, from 1983, when $11 million were appropriated, it has grown steadily to $43 million in 1989.

There is little national information available on the extent to which inservice programs—or other important professional resources—are used. Most of the existing data on this topic were collected from teachers, through self-reporting, in 1985-86 and reported in Weiss (1987). The SASS local education agency questionnaire asks whether the district reimburses teachers' tuition and course fees. It also asks whether free retraining is available for teachers for shortage areas, and what those shortage areas are. The school questionnaire for the 1990 follow-up of NELS:88 asks principals (primarily of middle or junior-high schools) whether teachers are rewarded with time off for professional workshops, extra materials, choice of classes, etc. Teachers in NELS:88 are asked about the number of hours spent on noncollege inservice education. The NEA Survey of the American Public School Teacher (described in Appendix B) includes three fairly detailed items concerning inservice of various types over the past three years, including how much of the teacher's own money was spent on college credit programs.

More data on policies related to inservice and other professional programs are needed from school districts. Among useful measures to obtain on inservice program use would be the number of hours of inservice training in mathematics, science, and related pedagogy accumulated in the last 12 months. Graduate courses should be distinguished from refresher workshops. Substantial inservice work in the form of graduate courses in one's primary field may indicate a high level of quality and professionalism or the intent to move from middle school to high school. The SASS teacher questionnaire asks whether in the past two years the teacher took any inservice or college courses requiring 30 or more hours of classroom study, the subject field, and a choice among several purposes for this continuing education. NEA's Survey of the American Public School Teacher, conducted every five years, also asks in considerable detail about kinds of inservice and college courses taken in various subjects.

Beyond survey data, in-depth interviews with personnel officers from a sample of SASS school districts, held on a regular basis in a conference format, are recommended. These conferences could yield information and context on inservice programs and the incentives behind them that no formal data collection can achieve. In this vein, improvements in inservice data are called for, perhaps using the NEA Survey of the American Public School Teacher as a starting point.

Other Practices That Affect Teaching Quality

This section discusses information on other practices affecting teaching quality that is relatively easy to obtain (and in some cases available in a national data set, as Appendix B shows).

Time allotted during the day for actual science and mathematics instruction. Darling-Hammond and Hudson (1987a:30) note some possible indicators of time use: (1) amount of time within the school day allocated to classroom instruction, preparation, nonteaching duties (bus duty, hall duty, etc.), meetings with colleagues, conferences with parents and students and (2) amount of time outside the school day teachers spend on planning and preparation, grading classroom assignments, contacting parents, working with students, completing administrative paperwork, reading professional journals, and participating in other professional development activities. With regard to the former category—time use during the school day—the National Research Council's Committee on Indicators of Precollege Science and Mathematics Education found that "the amount of time given to the study of a subject is consistently correlated with student performance as measured by achievement tests . . ." (National Research Council, 1985:106). Although it is possible to estimate instructional time through course enrollment at the secondary level, teachers at the elementary level have considerable latitude in the amount of time allocated to science and mathematics. Because of concern about the small amount of time spent in science instruction (Weiss, 1978), the committee recommended that time spent on science and mathematics instruction in elementary school be tracked on a sample basis at the national, state, and local levels (National Research Council, 1985:106-7).

Among current national data sets, the SASS and NELS:88 teacher questionnaires and the NSF survey of science and mathematics education appear to provide the most detailed data on time use.

Class size and teaching load bear on the teacher's ability to be effective. Darling-Hammond and Hudson (1987a:30-31) note some evidence that smaller class sizes are related to higher achievement, and they believe that "the relation between class size and teacher satisfaction and commitment has been, apparently, too obvious to warrant much study" (p.30). Such indicators should be monitored regularly, with particular attention to changes over time. The SASS, NELS:88, the NEA teacher questionnaires, and the NSF survey of science and mathematics education collect such data.

Opportunities for collaboration and decision making, when encouraged, lead to more discussion of teaching, more use of new ideas, more involvement in solving teaching problems, and stronger commitment to teaching (Darling-Hammond and Hudson, 1987a:32). Collaboration and participation in decision making also seem to reduce absenteeism and turnover. Schools vary widely in opportunities for collaboration. In a medium-sized urban district studied by the panel, new teachers received little support beyond being handed the syllabus for the course. One new science teacher said that she was sure that help was available but she had no time for discussions; she spent every free minute setting up or taking down labs. One school included in another district case study pairs each new science teacher with an experienced teacher, and interaction is frequent. The SASS and NAEP teacher questionnaires and the NSF survey of science and mathematics education collect general, self-reported opinions by teachers on these aspects of quality. While it would be difficult to measure the effects of these practices with survey instruments, they are noted because researchers have found them to be related to teaching quality.

Salaries do seem to affect individuals' decisions not to enter teaching, and low salaries influence existing teachers' propensities to take second jobs (Darling-Hammond and Hudson, 1987a:33). But research has shown little about the effects of salaries on teacher performance. Among the current national data sources, the National Education Association provides data on starting salary of teachers and the SASS teacher questionnaire inquires about salary, including income from nonschool employment and total family income. This questionnaire also requests opinions of various pay incentives.

A wide variety of data collection and research concerning school system policies and practices that affect teaching quality have been proposed in this section. We set priorities on these data needs in Chapter 6. It is critical, however, to build a foundation of data about school and district practices relating to quality and to embed the data in a context obtained by interaction with school districts (such as those recommended at the end of Chapter 6).

MEASURING TEACHER QUALIFICATIONS

A good information base on the quality of the teaching force would permit descriptive profiles of teachers and would allow for the measurement of change in teacher characteristics over time. Such information may be particularly important for understanding the quality adjustments that bring supply and demand into equilibrium. In the future, it may be possible to introduce quality information focusing on teachers, as well as on school and district practices, into models of teacher supply and demand.

Although certification is available as the baseline measure of teacher quality, it is a most imprecise indicator. More comprehensive information would be available from a teacher's transcript showing all courses taken. A higher standard of quality still would be approval by a professional board of standards in science or mathematics. Thus, one presumes a higher level of teacher knowledge as more of the standards are met and, therefore, a better quality of instruction. It is clear that the number of teachers meeting the standards declines as one moves from state certification to those of professional associations, and to the qualitative rather than the quantitative dimension of the professional standards. We discuss these standards of quality in order of difficulty of attainment.

Certification as the Basic Proxy for Teacher Quality

When measuring quality of the supply of teachers of science and mathematics, certification is the obvious first standard. Despite differences among states in certification rules and the level of preparation implied by the different standards, as shown in Appendix Table 5.1 (the tables in this chapter appear at the end of the chapter), certification is easily monitored. Certification does suggest some minimal level of knowledge and training.

Recently, alternative certification programs have been established in 21 states, in response primarily to shortages in particular subject areas, but also in response to dissatisfaction with the quality of traditional university programs (McKibbin, 1988). What do we know about the quality of teachers certificated through these programs? In a survey of these alternative programs, McKibbin concluded: "In most cases, the entry requirements were equal to or greater than the requirements for entry into university teacher education" (p. 34). But the weaknesses of the alternative programs (which supply a very small percentage of all new hires in the states that permit them) are similar to the weaknesses of traditional programs, McKibbin added: "In the larger programs the training resembles the offerings in university certification programs" (p. 35). He concluded that, although alternative programs thus far do meet specific subject-area needs, they do not seem to be superior in quality and they are not likely to replace conventional routes to certification" (p. 35).

Certification, whether obtained through traditional or alternative certification programs, is generally a poor indicator of quality, even for teachers of science and mathematics. Requirements are often nonspecific with respect to required breadth and depth of subject knowledge. For instance, in some states only a minimal number of hours in mathematics is specified with no level indicated (e.g., no requirement that the courses must contain work in calculus or beyond). Certification is also possible in a minor area, or as an endorsement on some other area; in these cases, even fewer hours of a given subject are required. Nonetheless, certification is the obvious first cut at a quality dimension of both the teaching force and the supply pool. Hiring uncertified teachers often means a diminution of quality. Conversely, increasing standards for certification can be expected to improve quality. Monitoring changes in the number of teachers with traditional, emergency, and alternative certificates, by subject areas, would provide useful information on quality at a baseline level.

Course Preparation and Transcript Data

What is known about the actual course preparation of teachers of mathematics and science? Two studies related to the question are (1) the analysis of mathematics and science preparation in the major teacher training institutions of the southern states reported by Galambos (1985) and (2) the study of a nationally representative sample of mathematics and science teachers, which includes their course backgrounds, reported by Weiss (1987). The research indicates that education majors tend to have less course work in mathematics and physical sciences work than do arts and science majors, although they tend to have more course work in biology and geology (Galambos, 1985). Education majors also have completed less college-level mathematics course work than have arts and science majors. The science preparation of education majors and arts and science majors is quite similar. Both groups take about the same number of science courses and accumulate about the same amount of laboratory experience. The groups differ in the relative amount of biology and geology versus chemistry and physics that they take: two-thirds of education majors take no chemistry or physics at all. Arts and science majors take almost twice as much chemistry and physics as teachers do.

Course preparation required to earn certification to teach physical sciences or life sciences can affect teacher quality in a rather unexpected way: by classifying geology and earth sciences (which are relatively descriptive courses) with physics and chemistry (which are quantitative), as is commonly done, a teaching candidate can obtain certification to teach physics/chemistry/geology by taking many geology/earth sciences courses and few physics and chemistry courses. The certificant can then be assigned to teach physics and chemistry classes with only limited knowledge of the subject matter. Disaggregation of geology from physics and chemistry in the science-teaching certification process should result in a closer fit between teachers' course preparation and their certification to teach certain courses.

Courses typically taken by those preparing to be elementary teachers and secondary teachers are described below in more detail.

Elementary Teachers

According to a recent RAND report (Darling-Hammond and Hudson, 1987b:32), the typical preparation program for elementary teachers includes four science courses and two and a half mathematics courses. A large proportion of elementary teacher preparation is in social science or general areas rather than in courses specifically related to subjects taught. Weiss indicates that elementary teachers are most likely to have taken a biological science course, a physical science course, but not chemistry or physics. These teachers are likely to have content-specific methods courses in both science and mathematics.[3]

Mathematical content courses designed specifically for elementary teachers help explain the increase among these teachers in their confidence to teach mathematics. Either they feel that they are well qualified, or their perception of elementary school mathematics is limited to arithmetic computation, which they feel comfortable teaching (Weiss, 1987). No such improvement in their confidence to teach science is noted. Either elementary teachers take too little science course work in college, or college-level science courses are not relevant for elementary teachers, or both.

The importance of high-quality instruction in science and mathematics at the critical elementary level cannot be overemphasized. Science in the elementary grades can become language arts—that is, vocabulary and not

[3]Mathematics for elementary school teachers in the Weiss data are courses developed in the 1970s and designed specifically for teachers. In many places they have solid course content that is more appropriate for elementary teachers than precalculus or introductory calculus courses. A consensus exists in the mathematics community on the concept of these specifically designed courses, which are considered to be a significant gain in elementary teacher preparation. Since these courses differ from the below-college-level courses in the Galambos' study, the two types of offerings should be distinguished from one another.

conceptual or experiential science, and the same can be said of mathematics. The question concerning the preparation of elementary teachers for mathematics and science gives rise to two contrasting answers. One answer calls for a better preparation of all elementary teachers in science and mathematics. The other answer calls for specialists to teach these subjects beginning at the fourth grade. One relevant issue is the youngest age at which children can comfortably tolerate varied teaching styles during the course of a day. While further research is needed toward raising the quality of elementary science and mathematics teaching, for the short run it is important to determine, perhaps through a future edition of SASS, the educational background of teachers at the elementary level. What science or mathematics courses or majors have they pursued?

Secondary Science and Mathematics Teachers

In the southern states included in the Galambos study, secondary teachers complete a major in their content areas. The Galambos study indicates, however, that they take fewer courses and fewer upper-division courses in their majors than do their counterparts in arts and sciences. Mathematics majors tend to resemble each other more closely than do science majors. Secondary teachers, like elementary teachers, tend to prepare more for biology than for chemistry or physics. As indicated above, this fact may result from widespread offerings or requirements for high school biology and more restricted offerings of chemistry and physics as electives. The cause and result are not clear. Availability of biology teachers may lead to a variety of high school course offerings. The Weiss data show that secondary science teachers do tend to concentrate in life sciences.

The Galambos and Weiss methodologies have limitations, and in some aspects their results are not comparable. The Weiss data on course background were self-reported. Galambos collected transcripts, but one does not know whether all the teachers trained in the Galambos study actually took teaching positions. However, the Galambos and Weiss studies do indicate the importance of course background data to gain a clear picture of actual preparation programs.

Data on courses taken and transcript data would seem the most concrete measure of qualifications. Differences in course titles among institutions and in course content could be assessed. Transcripts could be used to examine the teachers' majors while in college or graduate school and to examine the teacher's academic preparation in terms of specific courses taken. The transcripts could also be used to identify the teachers who failed, withdrew from, or repeated required courses in mathematics or science.

Working from such a disaggregated data base, one could determine, when a teacher does not meet certification requirements, whether it is a case of (1) lack of subject background, (2) lack of student teaching, (3) lack of content-specific methods courses, or (4) lack of general education courses. When a person meets certification requirements, one could also determine the degree to which the four factors considered above are satisfied. Transcripts are especially useful for assessing the backgrounds of recent graduates. They are also valuable for monitoring the changes taking place in teacher training programs. Teachers educated 10 years ago, for example, are unlikely to have had a course in computer science. Teachers trained today would be more likely to have that exposure. Transcripts would document that sort of change. Transcript data thus permit measurement of change over time in preparation programs, and they can suggest the prevalence of the four factors mentioned above among noncertified teachers.

Collecting transcript data for national purposes, however, could prove too daunting a task except for a relatively small sample. NSF has contracted for a transcript study for the science and mathematics teachers who responded to the teacher questionnaire in NELS:88. This appears to be an appropriate group of teachers to study since NELS:88 will also have student outcome data and data on teaching practices that can be analyzed in conjunction with the teacher background information.

When teachers move from elementary to secondary positions, do they take more science or mathematics courses to strengthen their content background? The SASS teacher questionnaire contains gaps in this area of information. Respondents note the number of courses they have taken in specific disciplines, but not when they were taken. One does not know whether they were taken before or after the teacher moved from elementary to secondary teaching, and therefore whether they were taken to strengthen background.

Professional Standards as a Quality Dimension

A measure of higher quality of the teacher supply would be the number of teachers meeting the professional standards of mathematics and science teacher associations for preparation programs. A slightly higher standard still would be teachers who also meet the inservice education standards of these associations. These professional association standards indicate whether teachers have a subject background in science or mathematics that includes a sense of how the discipline should be taught with regard to content and student background (Richardson-Koehler, 1987).

The recommendations of professional associations of mathematics and science teachers call for more than certification for measuring the quality of the teaching force. The abbreviated standards issued by the National Science Teachers Association (NSTA) and the National Council of Teachers of Mathematics (NCTM) indicate both the quantity and the pattern of preparation (Appendix Table 5.2). The standards cover both elementary and secondary teachers. Both the Galambos and Weiss studies suggest that teachers often meet the quantitative standards but may nevertheless fail to measure up to these criteria for quality: teachers take too many low-level courses, and they devote too little time to the quantitative physical sciences in favor of biological and descriptive sciences. For science teachers, the frequent mismatch between preparation and assignment leads to instructional situations in which the published professional standards are not met.

For secondary mathematics teachers, these standards are more likely to be breached in terms of the pattern and quality of preparation, not in terms of the total number of courses taken in mathematics, or the fact that one majored in mathematics. A common yardstick of these and other standards is that a secondary mathematics teacher's preparation should encompass more than introductory calculus. More importantly, the preparation program should sample areas within mathematics and culminate in an overall sense of the discipline.

Most state certification requirements for mathematics include a content-specific methods course. Some of the more recent alternative routes to certification do not have this requirement. Professional preparation specific to the teaching of mathematics, including understanding of mathematics learning, is an important dimension of quality. Both the NCTM and the Mathematical Association of America (MAA) guidelines include specific recommendations for mathematics education courses. Data on the extent to which these and inservice standards of professional associations are followed should be collected and monitored over time, building on monitoring activities conducted by the professional associations themselves.

Testing for Subject-Matter Knowledge

Most states now require that teachers pass a competency test as a prerequisite for certification (Appendix Table 5.3). By the fall of 1987, 45 states had enacted competency testing programs as part of the certification process. And in 31 states, rules also required that students take an examination for admission into a teacher education program.

One subject of debate concerns what competency tests should cover. No nationally accepted test exists, so some states use commercially developed tests, and others design their own. They cover a combination of basic skills, subject matter knowledge, and pedagogy. Appendix Table 5.3, which shows the states mandating competency testing of teachers, indicates the variety of tests employed. These range from low-level tests for screening entrance to teacher education programs to exit tests from these programs or for hiring, to more sophisticated measures such as the National Teachers Examinations (NTE) area tests for teachers within a field. The NTE tests for both subject matter and content-specific method. From test results, it does not appear that the ability of prospective teachers qualified in chemistry, physics, and general science has been declining from 1980 to 1984. However, as a measurement of instructional quality, the major limitation of the NTE test is the weak relationship of test performance to teacher performance (National Research Council, 1988:96). Other limitations of the NTE are the fact that not all states require the test and the fact that some test takers may not have taken teaching positions (National Science Foundation, 1985:127).

The NRC Committee on Indicators of Precollege Science and Mathematics Education recommended, as a key indicator of quality, "that samples of current teachers be selected to take tests that probe the same content and skills that their students are expected to master" (NRC, 1988:9-10). More specifically, the committee recommended that tests be given every four years to a sample of all teachers and every two years to a sample of newly hired secondary school science and mathematics teachers.

The Holmes and Carnegie Recommended Standards

Within the past three years, two groups of education experts have proposed more sophisticated measures of the quality of teachers' professional preparation. These groups call for placing greater requirements on teachers in the preservice stages—ensuring higher quality through more rigorous preparation, certification and selection—and ultimately for more professional autonomy once in the classroom. The two groups are (1) the Carnegie Task Force on Teaching as a Profession, which has given a grant to Stanford University to develop measures of teacher quality that may be used by its proposed National Board for Professional Teaching Standards (Carnegie Task Force on Teaching as a Profession, 1986) and (2) the Holmes Group, composed of 96 deans of education from universities nationwide, which aims to develop higher standards for teacher education at their institutions (Holmes Group, 1986).

The Carnegie group has proposed a three-stage voluntary assessment process covering subject matter mastery, education courses taken, and actual teaching performance, all under the aegis of a National Board for Professional Teaching Standards. Researchers at Stanford have classroom-tested measures of teacher quality for elementary school mathematics and high school history teachers. Both of these classroom-based studies of measures of quality seek to incorporate an evaluation of actual teaching effectiveness with regard to subject matter, content-specific teaching, and student characteristics. The development of these procedures could provide better measures of quality than those now available.

Both the Carnegie and Holmes recommendations are similar in calling for completion of a subject-matter major before initiation of training for teaching. Both permit an introduction to education, but only at the undergraduate level. Both are similar in calling for a restructuring of the teacher corps. (Appendix Table 5.4 summarizes their major recommendations.) The Holmes Group categorizes teachers as "career professionals," "professional teachers," and "instructors." The Carnegie Forum distinguishes between "licensure" and "certification." Licensure would be what is now called state certification. Beyond that, Carnegie's proposed National Board for Professional Teaching Standards would give board "certification." The assessment technique planned for board certification would go beyond knowledge and preparation of teachers to an assessment of their mastery of teaching techniques in the classroom. The ultimate measure of the quality of the teaching force would be the number of teachers that were board certified or that were categorized as "career professionals" under the Holmes definition.

The purpose of these measures is to enhance the general professionalism of the field and thereby attract and retain higher-quality personnel. As these approaches are refined and implemented more widely, more sophisticated measurements or data on aspects of teacher quality should emerge.

The changes called for by these groups, to the extent that they are adopted, will revamp many existing practices and give rise to new questions, some of which center on the supply response to changes in quality requirements. If teachers are required to study for five years, what adjustments are required in compensation to attract teachers? If teachers are required first to have a subject-matter major, will fewer or more persons continue on for a teaching degree? Will states and localities fund the change? Will the changes be willingly embraced or grudgingly made by prospective teachers, school administrators, school boards, and taxpayers?

The Presidential Awards for Science and Mathematics Teachers

The Presidential Award for Excellence program for recognition of excellence in teaching, sponsored by the National Science Foundation, was begun in 1983. Each year an award is given to one science teacher and one mathematics teacher in each state. Actually 54 jurisdictions are covered, consisting of the 50 states, the District of Columbia, Puerto Rico, the U.S. Trust Territories, and the Department of Defense dependency schools. Eligibility is restricted to teachers in junior high, middle, or high schools with a minimum of five years of experience who devote at least half-time to classroom teaching.

An examination of descriptions of winners reveals a profile of highly visible teachers. Many have published articles in professional journals, and all are involved heavily in after-school curricular activities, such as workshops and student projects. A few have higher degrees in their fields.

Each winner receives a monetary award of $5,000 given to the school. The candidate indicates how the money should be used. The range of uses includes travel expenses to attend courses, stipends for outside speakers, computer hardware and software, and science equipment. In a number of instances the money has been designated for materials that one would think the school budget should normally provide. There has been no follow-up on how the money is actually spent.

We recommend that there be a follow-up study at schools of previous winners to determine "quality" effects of the awards. In how many schools were the monies used for basic materials that would normally belong in the school or district budget? To what extent could the findings from follow-up analysis of the awards and recipients yield information about quality of instruction and the qualifications of the teachers?

The award winners constitute an interesting group for research. As an example of possible research, one study of 34 winners of the 1983 Presidential Award for Excellence in Teaching Mathematics was conducted by Yamashita (1987) to compare their level of professional development with that of a comparison group who were members of NCTM. A list of 21 professional development activities was given to all the participants to rate for importance to their own professional development. Awardees rated the most important activities as attending conferences and institutes, reading and writing for journals, developing curriculum beyond that for their immediate courses, advising student math activities, and teaching inservice courses. The comparison group rated writing for publication and consulting as of primary importance; the other activities mentioned above were rated less important to them than to the awardees. Awardees participated in more activities than did the comparison group. Yamashita concluded: "It may well be that the most distinguishing difference between the awardees and the comparison teachers in this study is the number of activities in which they engage and the higher energy level manifested therein" (p. 66).

TEACHER QUALIFICATIONS AND STUDENT OUTCOMES

Evidence

The literature to date does not indicate strong relationships of measurable teacher qualifications and such educational outcomes such as student performance on standardized tests. In a meta-analysis of 65 studies that had sought relationships between science teachers' characteristics and teaching effectiveness or student outcomes, Druva and Anderson (1983) find generally weak correlations, and many of these correlations were based on only one study. However, certain positive correlations are identified, on the basis of more than one study, that warrant statements of results: teaching effectiveness is positively correlated with the number of education courses taken, the student teaching grade, and length of teaching experience. Student outcomes are positively correlated with teachers' science training and general educational preparation. And this correlation between teachers' science training and cognitive student outcome is progressively higher in higher-level science courses.

From studies summarized in a comprehensive literature review by Darling-Hammond and Hudson (1986:24-32), it appears that certain teacher characteristics exhibit some positive relationship (often weak) to student performance: verbal ability; number of mathematics credits (for mathematics teachers); educational background in science, particularly for science teachers in higher grades; recent continuing educational experience; involvement in professional organizations; years of teaching experience; and positive attitudes toward teaching, flexibility, and enthusiasm. Other measures, such as IQ, National Teacher Examination (NTE) scores, and various measures of subject knowledge, have not shown any relationships to outcomes.

In a review of the literature, Blank and Raizen (1986) note that the failure of any research to establish a strong relation between teacher characteristics and student outcomes may be explained by a number of problems with the research to date on teacher effectiveness:

- The degree of variation in the independent variable, e.g., NTE scores, is often so small that no effect on outcomes would be measurable.
- Many studies have not included teachers with emergency certificates or low levels of training in the field in which they were teaching, so that, again, one would not expect to find strong relationships of such measures as extent of subject preparation and outcomes.
- Many studies have used student achievement tests as the sole measure of outcomes. The tests themselves may not relate to the goals of the students' courses; moreover, other measures such as attitudes toward science or math might show different results.

In sum, the presumption of a relationship between higher teacher qualifications and improved instruction is in need of testing and should not be discarded.

Implications for Data and Research

The panel reaffirms its earlier recommendations that relate to measuring teacher qualifications and their relationship to student outcomes (National Research Council, 1987c:8). Both the case studies and the meeting with large district personnel officers confirmed the usefulness of these recommended data collection efforts. As stated in the interim report:

1. We recommend that the National Center for Education Statistics surveys of teachers regularly include:
   - Measures of general intellectual ability and of academic preparation to teach mathematics and science fields, particularly for new entrants, in order to provide time series for monitoring and analysis. These measures should be obtained to the extent possible from transcript records rather than through survey questions.
   - For experienced teachers, measures of recent inservice preparation and participation in professional activities in mathematics and science fields. These surveys should also obtain measures of years of teaching mathematics and science distinct from total teaching experience.
   - Measures of certification (type and subject fields). We also recommend that the NCES obtain and disseminate available information on state certification policies and practices; we note that NCES has since published such information (NCES, 1988b, p. 123).
2. We recommend that further research be conducted on the relationship of measurable characteristics of teachers of mathematics and science to educational outcomes of students in these fields. In order to permit comprehensive and methodologically appropriate research on this issue, the National Educational Longitudinal Study of 1988 should include appropriate measures of student outcomes together with a rich set of teacher characteristics and characteristics of schools and districts. (We note that NCES includes such data items in NELS-88.)

Research relating teacher qualifications and student outcomes may be pursued using student and teacher questionnaire data from the 1985-86 NAEP assessment for science and mathematics as a starting point. NELS:88 is another useful source of information, especially for longitudinal research.

Of all the national data sets highlighted in Appendix B, the Schools and Staffing Survey asks for the greatest level of detail regarding teachers' qualifications. SASS asks teachers in detail about their past teaching experience, breaks in service, and previous occupation. It asks for the teacher's major and minor at every postsecondary level completed, the year each degree was completed, and the name of the undergraduate college. The respondent further notes in what field he or she teaches the most classes and the second-most classes. The respondent then is to provide the number of courses taken in these fields. For teachers who teach any science or mathematics in grades 7-12, there are data on number of courses in science and mathematics areas. SASS further asks detailed certification questions regarding field(s) and type (full, probationary, or emergency) for each field. The teacher then describes the amount—and purpose of—inservice or college courses taken in the last two years. Whether these courses were taken in the teacher's primary assignment field is also discernible through the questionnaire. Thus, with its ability to single out science and mathematics, the SASS teacher questionnaire will advance the level of statistical information on teacher qualifications far beyond the mere presence or absence of certification.

From the district's perspective, SASS asks district respondents which screening devices they use—or require—for hiring: full state certification; emergency certification; graduation from an approved teacher education program; college major or minor in the field to be taught; passing of a district test; passing of a state test of basic skills; passing of a state test of subject knowledge; passing of the NTE.

Screening devices used by districts, which constitute standards of qualification, and their changes over time should be monitored through SASS and by continued collection and dissemination of certification data from states. However, the collection and use of more statistics related to teacher quality must be tempered as their limitations are recognized.

There are some statistics not included in SASS, such as NTE scores, grade point averages, and other information that transcripts would provide. The NSF-sponsored teacher transcript study being carried out in conjunction with NELS:88 will provide the opportunity to explore the potential of transcripts as measures of academic background. In addition to transcript data, monitoring changes in admission standards for teacher education programs, by publishing data collected by the American Association of Colleges for Teacher Education (AACTE), is also recommended.

Finally, research is needed on the supply response to changes in certification requirements. By itself, more stringent qualification requirements will tend to reduce the supply of new teachers unless it is offset by salary prospects, greater prestige, or better working conditions. Absent any of these offsets, tougher qualification requirements are likely to shift supply between school districts and states, not produce a more qualified supply pool.

OTHER SCHOOL AND HOME FACTORS THAT AFFECT OUTCOMES

Much of the impetus for concern over the quality of precollege science and mathematics teachers arises from the widespread evidence that U.S. student outcomes—test scores and general level of literacy—in science and mathematics are poor. The possibility must be raised, however, that the problem underlying these low outcomes does not lie solely with the quality of teaching or the qualifications of teachers. Educational outcomes are a complex product of student and family inputs, teaching inputs, and educational curricula. Poor outcomes can be due to factors entirely beyond the quality of the teacher corps. This section addresses some of the most important of these factors.

Curriculum Structure

The influence of curriculum structure on U.S. students' mathematics test scores is under debate. It is argued that the consequences of a layered curriculum—through which students are introduced to relatively little new material each year through grade 8, and much of the mathematics training in any given year is thus basically review—are boredom and lack of mastery of the key ideas involved in the development of mathematical skills. A related criticism is that mathematics textbook producers, in trying to market their product to as many school systems as possible, end up with a light treatment of many topics rather than intensive treatment of a few topics. Since the basic text is the primary resource used by most precollege mathematics teachers (Weiss, 1987:31, 39), and since the text usually favors breadth and memorization of facts over depth (Office of Technology Assessment, 1988:30-34), the result is that students master few if any of the key concepts.

Quality of Textbooks

Although most science and mathematics teachers surveyed by Weiss in 1985 seemed to indicate that poor quality of textbooks was not a serious problem (Weiss, 1987:40-42), many scientists and educators who have reviewed the textbooks criticize their quality and their extensive use in classrooms (Office of Technology Assessment, 1988:30-33). The Mathematical Sciences Education Board (MSEB) of the National Research Council, as part of an ongoing effort to identify the key elements needed for reform, has determined that, between the second and eighth grades, there is only one year in which more than half the material is new (National Research Council, 1987b). This suggests that the solution to improving the quality of student skills in mathematics does not rest solely with providing better trained teachers, or even with providing more time for the teaching of science and mathematics, but rather depends on more fundamental reform of how the mathematics curriculum is organized. In a decentralized school system such as we have in the United States, fundamental reform of this nature is difficult to achieve. Moreover, there are differences of viewpoint among educators as to the validity of this line of criticism.

Classroom Time Used for Science and Mathematics

The amount of classroom time devoted to science and mathematics is another area of dispute, especially at the elementary level. Elementary school teachers are said to spend relatively little classroom time on science and mathematics topics. Recent studies have compared the amounts of classroom time spent by students on different topics; the evidence is mixed.

To begin with, there appears to be a substantial difference in the instructional time allocated to reading and mathematics in the early grades in the United States. One study indicates that about twice as much time is allocated to reading as to mathematics in the fourth grade (Cawelti and Adkisson, 1985). Weiss (1987:13) also found substantially more time devoted to reading than to mathematics, though not twice as much time. At the grade 4-6 level, teachers in this survey reported spending 63 minutes per day on reading and 52 minutes on mathematics. At the K-3 level, however, reading took up 77 minutes and mathematics 43 minutes, a wider difference in the earlier years of schooling. Forthcoming data from SASS will provide more recent information on classroom time; the SASS teacher questionnaire asks elementary school teachers in self-contained classes for hours per week spent in each of the core subjects, including science and mathematics.

Other studies based on careful observation of actual classroom time spent on mathematics in three cities (one each in the United States, Japan, and Taiwan) have found very large differences between the students in the U.S. city and those in the Taiwanese or Japanese city (Stevenson et al., 1986): U.S. fifth-grade children spent 3.4 hours per week on mathematics, Taiwanese students 11.7 hours per week, and Japanese students 7.8 hours per week. In grade 1, the differences were similar—2.7 hours for U.S. children, 4.0 hours for Taiwanese children, and 5.8 hours for Japanese students. In addition, U.S. students were less likely to be attending to the teachers than either Taiwanese or Japanese children, largely because individual work is much more common in U.S. classrooms than in Asian classrooms. However, for eighth grade, another study of classroom hours (McKnight et al., 1987) reports that U.S. students in grade 8 spend more time on mathematics instruction than students from Japan or Hong Kong.

Comparability problems limit attempts to draw conclusions from these studies. To start with, the studies are of students in different grades. In addition, the study by Stevenson and his colleagues contains very accurate measurement of classroom hours but covers a very small and possibly unrepresentative sample of schools; the McKnight et al. study is based on a national probability sample of schools but suffered from a high nonresponse rate and used officially scheduled hours and similar data to estimate time spent. Given the known difficulties of getting accurate estimates of time spent on various activities from very generalized methods (How much time is scheduled? How much is spent on average?), the panel is inclined to believe that the data of Stevenson and his colleagues are probably closer to the truth, and that one source of the difference in mathematics achievement is the gap in time allocated within the classroom.

If students' low skills and test scores in science and mathematics were known conclusively to be due simply to the relative amounts of time spent on these subjects, the solution would be relatively simple—provided school systems can be encouraged or induced to change the structure of their curricula. But if time spent or curriculum structure are the basic problems, then the issue is not one of teacher or teaching quality, but simply one of relative emphasis within the curriculum.

It may be true, of course, that many U.S. elementary school teachers are less comfortable teaching science and mathematics than teaching language arts, and therefore spend less time on science and mathematics. This may be explained by observing that in the United States elementary teachers tend not to be subject specialists, whereas the employment of specialist teachers of mathematics is more common in Japan, China, and Taiwan in the early grades.

Other Instructional Factors

Other possibly important differences between U.S. and Asian science and mathematics instruction have been identified in the ongoing studies being conducted by Stevenson and his colleagues. There are documented differences in the nature of textbooks—American texts explicate mathematics problems much more extensively and lead the students very carefully through exercises and problems; Asian texts are much shorter (about half the length in some of the texts examined) and make much stronger demands on the students to find their own way through the problem. There are also documented differences between American and Asian teachers of mathematics in the number of actual teaching hours per day and the amount of time available for planning and preparation; American teachers have much less nonteaching time scheduled during the day than their Asian counterparts. And there are documented differences in the degree to which teachers are autonomous in their own classrooms. In American classrooms, it is not uncommon that teachers are basically on their own after the first year, while in Asian classrooms younger teachers are typically under the tutelage of a senior teacher for a number of years (Stevenson, 1987; Lee et al., 1987; Stevenson et al., 1988; Stigler et al., 1987; Stevenson and Bartsch, in press). There also appear to be substantial differences in the training given to U.S. and Japanese mathematics teachers, with U.S. teachers spending more time learning mathematical content, and Japanese teachers spending more time learning mathematics pedagogy (teaching of mathematics) (McKnight et al., 1987:65).

Home Environment

The home environment also has a significant impact on young children's learning. The home environment of many children in the United States is not conducive to concentrated thought and learning. The proportion of single-parent households and the proportion of households in which both spouses work are much higher now than in past decades. These realities can create problems for children and can have a potentially serious influence on their skill development. To understand educational outcomes requires us to understand the contributing effects of these home environment factors.

Parental attitudes, as well as demographic differences in home environments, can also influence children's ambition to concentrate on academic learning. The best documented evidence of differences in attitudes and expectations comes from a comparison of American and Asian households. In general, Asian mothers are less satisfied with the school performance of their children than American mothers (despite the fact that their children are generally doing better); they are more likely to attribute success in school to hard work rather than to native ability; and they are less likely to be satisfied with the way the schools are performing than their American counterparts (Lee et al., 1987).

Poor student outcomes are thus not uniquely correlated with inadequate quantity or quality of teachers, but could easily be due to factors that are largely unrelated to teacher quality. One cannot conclude that poor science and mathematics outcomes on the part of students necessarily reflect inadequacies in the background or ability of their teachers and to try to remedy the problem only by enhancing the numbers or the quality of precollege science and mathematics teachers. Factors such as the structure of the curriculum, the practices of both K-12 school systems and teacher training institutions, the amount of time spent on science and mathematics topics in schools, and the influence of home environments on development outcomes all need to be understood before we can fully understand the problem or devise appropriate remedies. Thus, although the issues raised in this section are beyond the scope of this study, they serve to point up the many other factors beyond teaching and teacher quality that bear upon student outcomes.

SUMMARY

In the near term, it is through quality adjustments that the supply and demand for precollege mathematics and science teachers reach equilibrium. The quality of instruction is therefore a central focus of our study. Statistics that can furnish indications of quality and trends or events that can be monitored to illuminate quality and changes in quality over time are called for.

Areas of data concern relating to quality not only focus on teacher characteristics but also extend to contextual arrangements that affect overall teaching quality. These contextual variables include teacher policies and practices regarding assignments and teacher background, course offerings and enrollments, recruitment practices, school and school system policies governing both initial placement and transfers, and inservice training provided by schools, school systems, states, professional associations, and the federal government. Also of concern is the distribution of qualified teachers across districts when classified by enrollment size; by racial/ethnic characteristics of its students; by geographic characteristics of urban, suburban, rural; and by socioeconomic status.

We have identified some district policies and practices that influence teaching quality, and note in particular the importance of information on recruitment practices, seniority rules, potential for teacher advancement, teacher assignment and misassignment, and continuing professional development, as well as external factors, primarily state mandates and policies, that affect the quality of the supply pool.

There are numerous ways to measure and assess teacher qualifications that influence overall teaching quality. Some are objective and can be counted; some are subjective and not easily quantified. Some are easily quantified but of little use (such as certification); some would be highly useful but would require more examination (such as transcripts). Some indicators are based on existing standards (such as those of the National Science Teachers Association), and some on proposed standards (such as those of the Holmes and Carnegie groups). While it is acknowledged that a thorough knowledge of content is only a necessary and not a sufficient set of characteristics for a successful teacher, certain qualifications are necessary. And data can be collected to indicate the presence and strength of these qualifications. We do recognize, however, the considerable amount of effort and resources that would have to be invested in collecting these data, when such factors as presence of certification, transcript data, and educational background have not yet been demonstrated to be strongly associated with teacher quality or student outcome. Thus, it is important that the National Science Foundation fund a program of controlled experiments on factors that do measure teacher or teaching quality. Such research would include identifying the relationship between measurable teacher qualifications and student outcomes.

If the Carnegie or Holmes recommendations for higher professional standards are adopted, the consequent changes in the teaching force should be monitored, together with any changes in supply as a result of the more rigorous requirements.

Other factors beyond teacher quality—such as textbook use, time commitments, the structure of science and mathematics curricula, and home environment—were noted as influences on teaching quality and student outcomes. These factors complicate any attempts to link outcomes with particular teacher qualifications.

In conclusion, to understand the crucial role of quality in bringing supply and demand for precollege science and mathematics teachers into equilibrium in the short term, we have acknowledged some rather daunting data needs and research issues. We realize that these needs might not be able to be met completely enough to introduce teacher quality measures into teacher supply models in the near future. But successful collection of more precise data, particularly through SASS and existing state information files, can be expected to contribute to an understanding of teacher quality, and additional research may help identify the characteristics of teachers and teaching that are determinants of student outcomes.

APPENDIX TABLE 5.1 Mathematics and Science Teacher Certification Requirements for Secondary School Teachers, by State, June 1987

	Course Credits by Certification Field						
	Mathematics	Science, Broad field	Biology, Chemistry, Physics	Earth Science	General Science	Teaching Methods: Science/ Mathematics	Supervise Teaching Experience
Alabama	27	52	27	27	27	Yes	9
Alaska	None	None	None	None	None	No	None
Arizona	30	30	30	30	30	Yes	8
Arkansas	21	-	24	24	24	No	12 wks
California	45	-	45	-	-	No	a
Colorado	b	b	b	b	b	Yes	400 hrs
Connecticut	18	-	18	18	21	No	6
Delaware	30	-	39-45	39	36	Yes	6
District of Columbia	27	30	30	30	30	Yes	1 sem
Florida	21	-	20	20	20	Yes(S)	6
Georgia	60 qtr	45 qtr	40 qtr	40 qtr	-	Yes(M)	15 qtr
Guam	18	18	-	-	-	No	None

APPENDIX TABLE 5.1 Mathematics and Science Teacher Certification Requirements for Secondary School Teachers, by State, June 1987 - continued

	Course Credits by Certification Field						
	Mathematics	Science, Broad field	Biology, Chemistry, Physics	Earth Science	General Science	Teaching Methods: Science/ Mathematics	Supervise Teaching Experience
Hawaii	b	-	b	b	b	b	b
Idaho	20	45	20	20	-	No	6
Illinois	24	32	24	24	-	Yes	5
Indiana	36	-	36	36	36	Yes	9 wks
Iowa	24	24	24	24	24	Yes	Yes
Kansas	b	b	b	b	b	b	b
Kentucky	30	48	30	30	-	No	9-12
Louisiana	20	-	20	20	32	No	9
Maine	18	18	-	-	-	Yes	6
Maryland	24	36	24	24	36	Yes	6
Massachusetts	36	36	36	36	36	Yes	300 hrs
Michigan	30	36	30	30	-	No	6
Minnesota	c	c	c	c	c	c	1 qtr
Mississippi	24	-	32	32	32	Yes(S)	6

Missouri	30	30	20	20	20	Yes	8
Montana	30 qtr	60 qtr	30 qtr	30 qtr	30 qtr	Yes	10 wks
Nebraska	30	45	24	24	-	Yes	320 hrs
Nevada	16	36	16	16	16	No	8
New Hampshire	b	b	b	b	b	b	b
New Jersey	30	30	30	30	30	No	c
New Mexico	24	24	24	24	24	Yes	6
New York	24	-	36	36	36	No	Yes
North Carolina	c	c	c	c	c	Yes	6
North Dakota	c	c	c	c	c	Yes	8
Ohio	30	60	30	30	30	Yes	a
Oklahoma	40	40	40	40	40	No	12 wks
Oregon	21	45	45	45	45	Yes(M)	15 qtr
Pennsylvania	b	b	b	b	b	b	b
Puerto Rico	30	30	30	-	30	Yes	3(S)5(M)
Rhode Island	30	30	30	-	30	Yes	6
South Carolina	24	30	12	-	18	Yes(M)	6
South Dakota	18	21	12	12	18	No	6
Tennessee	36 qtr	48 qtr	24 qtr	24 qtr	24 qtr	Yes	4
Texas	24	48	24	24	-	No	6
Utah	c	c	c	c	c	Yes	12
Vermont	18	18	18	18	18	Yes	None
Virginia	27	-	24	24	30	No	6
Virgin Islands	24	NA	NA	NA	NA	No	Yes

APPENDIX TABLE 5.1 Mathematics and Science Teacher Certification Requirements for Secondary School Teachers, by State, June 1987 - continued

	Course Credits by Certification Field						
	Mathematics	Science, Broad field	Biology, Chemistry, Physics	Earth Science	General Science	Teaching Methods: Science/ Mathematics	Supervise Teaching Experience
Washington	24	51	24	24	-	No	15
West Virginia	c	c	c	c	c	c	c
Wisconsin	34	54	34	34	34	Yes	5
Wyoming	24	30	12	12	12	No	1 course

Key: Course credits = semester credit hours, unless otherwise specified; qtr = quarter credit hours; M = mathematics only; S = science only; NA = not available; blank space = no certification offered.

[a] 1 semester full time or 2 semesters half time--California; supervised teaching experience and 300 hours clinical/field-based experience--Ohio.

[b] Certification requirements determined by degree-granting institution or approved competency-based program.

[c] Major or minor--North Dakota, Utah; 20 to 40 percent of program--Minnesota and North Carolina; courses matched with job requirements--West Virginia.

Source: Office of Technology Assessment (1988:59).

APPENDIX TABLE 5.2 Guidelines for Mathematics and Science Teacher Qualifications Specified by the National Council of Teachers of Mathematics (NCTM) and the National Science Teachers Association (NSTA)

NCTM Guidelines	NSTA Standards
Early elementary school	Elementary level
The following 3, each of which presumes a prerequisite of 2 years of high school algebra and 1 year of geometry: 1. number systems 2. informal geometry 3. mathematics teaching methods	1. Minimum 12 semester hours in laboratory- or field-oriented science including courses in biological, physical, and earth sciences. These courses should provide science content that is applicable to elementary classrooms. 2. Minimum of 1 course in elementary science methods (approximately 3 semester hours) to be taken after completion of content courses. 3. Field experience in teaching science to elementary students.
Upper elementary and middle school	Middle/junior high school level
The following 4 courses, each of which presumes a prerequisite of 2 years of high school algebra and 1 year of geometry: 1. number systems 2. informal geometry 3. topics in mathematics (including real number systems, probability and statistics, coordinate geometry, and number theory) 4. mathematics methods	1. Minimum 36 semester hours of science instruction with at least 9 hours in each of biological or earth science, physical science, and earth/space science. Remaining 9 hours should be science electives. 2. Minimum of 9 semester hours in support areas of mathematics and computer science. 3. A science methods course designed for the middle school level. 4. Observation and field experience with early adolescent science classes.
Junior high school	Secondary level
The following 7 courses, each with a prerequisite of 3 to 4 years of high school mathematics, beginning with algebra and including trigonometry: 1. calculus 2. geometry 3. computer science 4. abstract algebra 5. mathematics applications 6. probability and statistics 7. mathematics methods	General standards for all science specialization areas: 1. Minimum 50 semester hours of course work in 1 or more sciences, plus study in related fields of mathematics, statistics, and computer applications. 2. Three- to 5-semester-hour course in science methods and curriculum. 3. Field experiences in secondary science classrooms at more than 1 grade level or more than 1 science area.

(Appendix Table 5.2, continued)

NCTM Guidelines	NSTA Standards
Senior high school	Specialized standards
The following 13 courses, which constitute an undergraduate major in mathematics, each presume a prerequisite of 3 to 4 years of high school mathematics, beginning with algebra and including trigomometry: 1-3. 3 semesters of calculus 4. computer science 5-6. linear and abstract algebra 7. geometry 8. probability and statistics 9-12. 1 course each in: mathematics methods, mathematics applications, selected topics, and the history of mathematics 13. at least 1 additional mathematics elective course Specialized standards	1. Biology: minimum 32 semester hours of biology plus 16 semester hours in other sciences. 2. Chemistry: minimum 32 semester hours of chemistry plus 16 semester hours in other sciences. 3. Earth/space science: minimum 32 semester hours of earth/space science, specializing in one area (astronomy, geology, meteorology, or oceanography), plus 16 semester hours in other sciences. 4. General science: 8 semester hours each in biology, chemistry, physics, earth/space science, and applications of science in society. Twelve hours in any 1 area, plus mathematics to at least the precalculus level. 5. Physical science: 24 semester hours in chemistry, physics, and applications to society, plus 24 semester hours in earth/space science; also an introductory biology course. 6. Physics: 32 semester hours in physics, plus 16 in other sciences.

Source: Office of Technology Assessment (1988:64).

APPENDIX TABLE 5.3 States That Have Enacted Testing Programs for Initially Certifying Teachers: Fall 1987

State	Enacted	Effective	Test Used[a]
Alabama	1980	1981	State
Arizona	1980	1980	State
Arkansas	1979	1983	NTE
California	1981	1982	CBEST
Colorado	1981	1983	CAT
Connecticut	1982	1985	State
Delaware	1982	1983	PPST
Florida	1978	1980	State
Georgia	1975	1980	State
Hawaii	1986	1986	NTE
Idaho	1987	1988	NTE
Illinois	1985	1988	State
Indiana	1984	1985	NTE
Kansas	1984	1986	NTE and PPST
Kentucky	1984	1985	NTE
Louisiana	1977	1978	NTE
Maine	1984	1988	NTE
Maryland	1986	1986	NTE
Massachusetts	1985	b	b
Michigan	1986	1991	b
Minnesota	1986	1988	PPST
Mississippi	1975	1977	NTE
Missouri	1985	1988	b
Montana	1985	1986	NTE
Nebraska	1984	1989	b
Nevada	1984	1989	PPST and State
New Hampshire	1984	1985	PPST and NTE
New Jersey	1984	1985	NTE
New Mexico	1981	1983	NTE
New York	1980	1984	NTE
North Carolina	1964	1964	NTE
North Dakota	1986	b	b
Ohio	1986	1987	NTE
Oklahoma	1980	1982	State
Oregon	1984	1985	CBEST
Pennsylvania	1985	1987	State
Rhode Island	1985	1986	NTE
South Carolina	1979	1982	NTE and State
South Dakota	1985	1986	NTE
Tennessee	1980	1981	NTE

APPENDIX TABLE 5.3 Continued

State	Enacted	Effective	Test Used[a]
Texas	1981	1986	State
Virginia	1979	1980	NTE
Washington	1984	b	b
West Virginia	1982	1985	State
Wisconsin	1986	1990	b

[a] Tests:
CAT = California Achievement Test;
CBEST = California Basic Skills Test;
NTE = National Teacher Examination;
PPST = Pre-Professional Skills Test;
State = State-developed test.
[b] To be determined.
Source: National Center for Education Statistics (1988f:249-250).

APPENDIX TABLE 5.4 Comparison of Recommendations of Carnegie and Holmes Reports Pertaining to Preservice Education of Teachers

Category of Recommendation	Carnegie Report [a]	Holmes Group [b]
Fifth Year of Study	Require bachelors degree in the arts and sciences as prerequisite of professional study of teaching. Require a master's degree for all teachers.	Make education of teachers more solid intellectually by pursuing an undergraduate major in an academic subject other than education, receive their professional training in a fifth year master's degree program, and complete a year-long supervised internship.
Curriculum Revision	Develop new professional curriculum in graduate schools of education leading to Master in Teaching degree based on systematic knowledge of teaching and including internships and residencies in schools.	Revise undergraduate curriculum in arts and sciences. Organize academic course requirements, including involvement of other departments in institutions of higher education. Need advanced studies inpedagogy (focus on human cognition, teaching and learning, and teaching), teachers' learning, assessment of professional performance, and evaluation of instruction.
Coordination	Connect institutions of higher education with schools through the development of professional development schools.	Need coherent program in schools and institutions of higher education that will support advanced study. Create professional development schools, similar to teaching hospitals, in which prospective teachers would receive their clinical training.
Certification	Create a national board for professional teaching standards to establish high standards for what teachers need to know and to be able to do, and to certify teachers who meet that standard.	Create 3-tier systems of teacher licensing: o Instructor--has BA degree, without year of supervised practice and study in pedagogy and human learning; has passed exams (see evaluation) o Professional teacher--has MA in teaching; completed year of supervised practice; passed exams o Career professional--has completed all of the above plus further specialized study

APPENDIX TABLE 5.4, continued

Category of Recommendation	Carnegie Report [a]	Holmes Group [b]
Evaluation/ Assessment		Use multiple evaluations o Test basic mastery of writing and speaking o Demonstrate mastery of subject, skill in lesson planning, and instructional delivery prior to clinical internship o Evaluate variety of teaching styles during internship--including own--and present analytic evidence as part of professional portfolio for advancement
Differential Staffing	Restructure teaching force and introduce new category of lead teachers with proven ability to provide active leadership in redesign of schools and in helping colleagues to uphold high standards of learning and teaching.	Recognize differences in teacher's knowledge, skill, and commitment in their education, certification, and work.

[a] Carnegie Task Force on Teaching as a Profession (1986) A Nation Prepared: Teachers for the 21st Century. Washington, D.C.: Carnegie Forum on Education and the Economy. Pp. 55-56.

[b] The Holmes Group (1986) Tomorrow's Teachers: A Report of the Holmes Group. East Lansing: The Holmes Group, Inc. Pp. 65-66.

Source: Regional Laboratory for Educational Improvement of the Northeast and Islands (1987:15-17).

6
Data Needs and Research Opportunities

The previous chapters display some of the conceptual richness that meaningful descriptions of supply and demand for precollege science and mathematics teachers would entail. Some of these discussions are based on quite detailed data from a single state or sample survey. Such rich descriptions cannot be realized at a national level, however, in the absence of comprehensive national data (or an aggregation of state data) that would give them substance.

As we have noted in earlier chapters, classrooms are rarely unstaffed. What usually is adjusted in times of shortage or surplus is the quality of staff. Moreover, there is a reservoir of individuals who are certified to teach, as well as individuals who would like to teach but are not certified, that is far larger than the counts generated by enumeration of those currently teaching and the numbers of new graduates of teacher training programs.

Some analysts believe that in the next decade the demand for science and mathematics teachers will increase as a result of growing enrollment demand and teacher retirements. To understand how the supply and quality of science and mathematics teachers will respond to changes in demand, data are needed to support the construction of measures of demand, of potential supply, of quality to the extent possible, and of models of the responsiveness of supply to incentives and to changes in demand.

It is the panel's view that current national data collection efforts and knowledge of the relation between incentives and supply are inadequate to support rich structural modeling of teacher demand and supply. Thus, we propose a sequential approach:

- First, as efforts are made to improve the consistency, scope, and quantity of data, publish indicators from existing data that are considered relevant to teacher supply and demand.
- Second, carry out the research needed to support behavioral models.
- Third, as data bases are improved and research findings on the relation between incentive and supply become available, devote resources to structural modeling that goes beyond straightforward extrapolative projection.

In this chapter, the panel's recommendations for data to monitor the state of demand, supply, and quality of precollege science and mathematics teachers are set out, followed by a discussion of research issues and a recommendation for a series of conferences that could assist in understanding the processes that result in the observed data. Some of the data recommendations could be easily implemented by modification of existing survey questionnaires. Others would require new data collection approaches. Similarly, some of the research issues can be studied with existing data or data soon to be available from SASS, whereas other issues can only be investigated by development of new data bases.

This chapter proceeds with sections on data recommendations (most of them for NCES), research issues (primarily for NSF), research facilitation suggestions, and finally a major recommendation for a series of conferences bringing together NCES and officials of school districts and state education agencies to discuss teacher supply, demand, and quality concerns. It is difficult to assign priorities across such disparate topics. However, within each section where we have listed specific recommendations we have marked those of highest priority with an asterisk. In addition, in most of the sections, the specific recommendations are listed in order of priority. Finally, at the conclusion of the Summary we offer guidelines for timing the implementation of the high-priority recommendations.

DATA RECOMMENDATIONS

We present a wide range of data recommendations related to demand, supply, and quality. Better data in the short run on elements of the supply/demand situation, including the sensitivity of teachers' career decisions to the many factors that may influence these decisions, will in turn contribute to the kinds of behavioral models that should be effective in the future. At the outset, we urge NCES to support SASS with a reliable, ongoing base of funding.

Funding for Data Collection

In the near term it is of key importance to monitor the state of demand, supply, and quality of precollege science and mathematics teachers. We note that the National Center for Education Statistics recognized the need for a major effort of data collection concerning teachers and contracted with the RAND Corporation in 1985 to design the Schools and Staffing Survey (SASS). A pilot test of that survey was conducted in 1986-87, and the first full-scale survey was conducted in 1987-88. SASS data are expected to be available in 1990 (publication has been delayed because of recent legislation pertaining to the confidentiality of data collected by NCES). Although new surveys always have unexpected problems, we anticipate that much of the needed data will be available and we have so identified data needs that SASS was designed to fill. To monitor the supply and demand of science and mathematics teachers effectively, SASS should be repeated periodically, at least every four years, and adequate funds for analysis should be made available to permit full exploitation of this valuable data resource.

**The panel recommends that provision be made in the budget for the National Center for Education Statistics to conduct the Schools and Staffing Survey on a regular cycle and that the budget include funds for follow-up surveys of teachers who leave teaching and for in-house and external analysis of the survey data.*

Data Related to Demand

In general, the panel finds demand data to be relatively adequate. The task of projecting enrollment-driven demand for science and mathematics teachers is relatively straightforward. The U.S. Bureau of the Census collects data on births and their geographic distribution. The children born each year move through precollege schooling in a very predictable way and are augmented primarily through immigration. The data most needed for projecting demand for teachers are current attrition data, particularly data on attrition for reasons other than retirement. Forecasting demand for science and math teachers, rather than teachers generally, could be improved if better data on course-taking behavior in high school were available. This behavior results both from state-mandated course requirements and from student course preferences. *The panel recommends the collection of additional data, disaggregated by subject, of the following types, in order of priority*:

*1. *Data on attrition/retention rates of nonretirees by discipline.* Detailed discussion of these data on attrition/retention is found below under data related to supply. Although knowledge and ability to forecast retirements is fairly adequate, data on attrition earlier in the teaching career would improve projections of demand.

2. *Data on state-mandated high school course requirements in science and mathematics.* Increased requirements for mathematics or science courses can lead to greater demand for teachers of both advanced and remedial courses. The Education Commission of the States (ECS) publishes data on changes in state requirements periodically, although at a very general level, i.e., the number of science and mathematics courses required. State data are only a beginning. NCES has found that high school graduation requirements as mandated by states are often exceeded by the requirements already in place in individual districts. Therefore, relying solely on changes in state requirements to determine demand for teachers will probably overestimate increased demand.

3. *Data on course offerings and changes in offerings over time.* Changes in course offerings can change the demand for science and mathematics teachers and, in particular, can indicate the need for teachers with special qualifications, such as the ability to teach advanced placement chemistry.

4. *Data on changes in enrollments (in general and for particular science and mathematics courses). Such enrollment data should be disaggregated by sex and race/ethnicity.* Although SASS data from LEAs will provide changes in the number of secondary science teachers by science subject and changes in the number of mathematics teachers and computer-science teachers, the SASS local education agency form does not track student enrollments in a parallel fashion. The SASS teacher questionnaire could be used to obtain enrollment data by race/ethnicity and sex for particular science and mathematics courses by expanding the question that asks teachers sampled for the names of the courses they teach and the number of students in each course.

It is important that new data collections related to demand be disaggregated by subject to be useful in setting policy to produce a corps of teachers with the right mix of skills to meet the demands of future years.

Data Related to Supply

The number of teachers employed in schools nationwide is augmented each year by new graduates from teacher training programs, newly certified teachers who enter teaching from other pathways (collaborative relationships with industry, for example), and entrants from the reserve pool of previously certified teachers who have never taught or former teachers who have chosen to reenter the profession. The number of teachers employed is diminished by attrition due to retirement and other causes. Thus monitoring supply requires keeping track of changes in the supply pool generally. In particular, it requires data over time on certification, on attrition and retention rates, and on new hiring and the levels of key incentives that attract people to the teaching profession. Monitoring supply calls, as well, for data that describe the competitiveness of teacher salaries relative to opportunity cost salaries, amount of reciprocity in certification across states, and portability of teacher pensions. Our highest-priority data recommendations for the short run call for better data on attrition, on the hiring rate for newly certified teachers, on the supply potential of segments of the reserve pool, and on incentives that influence individuals' decisions to enter or leave teaching. *The panel recommends obtaining better data or making fuller use of existing data on the following aspects of supply, which are presented in "pipeline" order*:

1. *College students planning to teach*

• *Trend data on the career interests expressed by college freshmen.* Indicators of future teacher supply include student aspirations to become teachers. The proportion of freshmen aspiring to teach appears to be up now for the first time in many years. Data from The American Freshman (described in Appendix B) should be analyzed by subject major, sex, and ethnicity. Follow-ups of students after two years and four years that are conducted occasionally also should be analyzed. However, it should be kept in mind that the number of freshmen who say they want to teach may be only loosely related to the number among them who actually obtain certificates.

• *Data on science, mathematics, and education majors to be related, if possible, to numbers of actual certificants.* How many, by major, who planned/did not plan to teach entered/did not enter the certification stage? The High School and Beyond and Recent College Graduates surveys provide these data.

2. *Certification*

• *Information on state certification policies and practices.* NCES should continue to collect and disseminate this information.

• *Data from states on education school and certification program enrollments* by subject specialty, sex, and race/ethnic group. NCES should collect and disseminate these indicators of what is in the pipeline–potential additions to teacher supply in the next one or two years.

• *National data on the number of new certificants by type (traditional, emergency, or alternative program) and by subject annually* compiled from state certification board data. Different states have different certification practices and categories. It should be possible for NCES to get comparable totals, however, for aggregated categories (mathematics, science, elementary, secondary, for example) and to present disaggregated data when available. Since teachers may be certified in more than one category, these data will not be perfectly matched with the increase in the supply of newly qualified teachers. Rather, these data provide an upper bound on the change in newly qualified supply.

• *Information on the degree of reciprocity in certification across states.* These data help to indicate the extent to which shortages in one part of the country could be filled by additional teachers from elsewhere. The National Association of State Directors of Teacher Education and Certification (NASDTEC) publishes states' reciprocity provisions periodically in its *Manual on Certification.* It would be useful to include this information in NCES's *Digest of Education Statistics.* The effect of reciprocity on the mobility of the reserve pool is an issue for research.

• *Data collected at the state or local level on the extent to which teachers are employed who hold temporary, provisional, or emergency certificates.* The SASS teacher questionnaire will collect this information. The data should be analyzed by subject, by region, or by type of area (e.g., rural, suburban, urban, or central city). These data may indicate exhaustion of the reserve pool or a shortage in a particular subject or in a particular geographic area. SASS also asks districts for the number of full-time equivalent positions, by subject, that remained vacant or were filled by a substitute or withdrawn for lack of a suitable candidate. SASS collects the total for these three categories but does not provide disaggregated data. Information on the reasons for employing teachers with these categories of certificates could be obtained through in-depth discussions with school district officials on a regular basis, as the panel recommends at the end of this chapter.

• *Data on the use of alternative programs for earning certification to teach.* In response to perceived shortages in the quantity or quality of teachers in general, or in some cases of teachers of particular subjects, a growing number of initiatives providing alternative or nontraditional routes to certification have recently been created. The extent to which science and mathematics teachers obtained their certification through an alternative program and the distribution of such teachers geographically and among urban, suburban, and rural schools should be monitored closely, through a question that can readily be added to SASS.

• *Follow-up data on new certificants to ascertain the numbers and proportions of new certificants who did not immediately take teaching positions.* Such data would probe for reasons behind their decisions, alternative activities chosen, and salaries, if possible. The Survey of Recent College Graduates is one possible source of data, as is its successor, the Baccalaureate and Beyond Longitudinal Study, which is expected to provide data in the future.

3. *New hires and incentives to teach*

*• *Comparative salary data* to indicate competitiveness of teachers' salaries relative to those of alternative nonteaching positions. Although there is a question of just how this comparison should be made, one simple

measure would be starting salaries in industry for people with equivalent education (e.g., a B.S. in mathematics). The College Placement Council (1988) publishes these data annually.

*• *Data on reasons why teachers selected their current school/district* and alternative offers they had. Such data (not currently collected by the SASS teacher surveys or the NLS teaching supplement) would help identify actions that schools or districts might take to attract well-qualified teachers. (Note that, if there is a national shortage, it is not clear that such actions would increase the national supply of teachers rather than the attractiveness of a particular school/district.)

*• *The number of last year's certificants, by type of certificate, who were hired (or received a firm job offer) by school district and the proportion of those who applied for positions and were hired.* These data, which can be collected from college placement offices, provide an indicator of the extent to which school districts draw from the pool of newly certified teachers, rather than teachers from the reserve pool. Inclusion of racial/ethnic data would help monitor the progress of minorities through the pipeline.

• *Trend data obtained from districts on the ratio of the number of applicants to vacancies in teaching, by field, and on the ratio of job offers per hire, by field.* Questions on the number of applicants and the number of job offers per vacancy could be added to the SASS survey. Although an applicant may apply for more than one vacancy, a decline in this indicator (assuming no change in recruitment practices) would point to increasing shortages of applicants in a particular field. Similarly, an increase in the number of job offers per vacancy could indicate a shortage or the need to make the positions more attractive.

• *Data from school districts (building on SASS) on the extent to which districts are shifting from screening applicants to recruiting, disaggregated by subject and by race/ethnic group.* When teachers are in surplus, districts recruit near home (if at all) and passively accept applications that are then screened. As shortages arise, districts recruit more vigorously. Thus, if it were possible to count the number of districts engaging in active recruiting without significant measurement error, it would be an indicator of shortage, and growth in such an indicator over time would indicate increasing shortage. Current SASS school district questionnaires ask about screening generally, but not by subject, nor do they ask about recruiting. Because of the difficulty of quantifying recruiting, this topic should be explored more thoroughly by in-depth discussions with a sample of school districts.

• *Recruitment data from personnel directors of school systems and from college and university placement directors that identify fields of shortage as they perceive them.* Such data could highlight teaching fields for which normal supply is not adequate. Expansion of recruitment areas and changes in practices such as early offers or bonuses might indicate the severity of the prospective shortage. Widespread reporting by many personnel officers and many placement officers of the need for teachers in a field might indicate a potential area of shortage. However, such a need would have to persist for several years before being classified as a shortage. Recruitment data should be collected by the geographic area covered, by subject, and by the race/ethnicity of recruits.

4. *The reserve pool*

Because the reserve pool is such a major source of new hires, it is important to know not only how large it is, but the content and size of its various components, its size in different states or regions of a state, and whether it is nearly exhausted in the area of the relevant labor market. From the panel's viewpoint, all the approaches under this topic are high priority because so much hiring is from the reserve pool. Knowing the characteristics of the types of people in the reserve pool is important, for some individuals would not reenter teaching under any conditions. Different components of the reserve pool can be expected to behave differently.

A variety of approaches could be taken to measure the supply potential of segments of the reserve pool:

*• *Follow new college graduates over time* to determine the proportion that enter teaching by the number of years after graduation, reasons for leaving teaching, time spent out of teaching, and reentry into teaching. Data from the longitudinal studies High School and Beyond and NELS:88 provide opportunities for studying the reserve pool from this perspective. Data on teaching status one year after graduation are available from the Recent College Graduates (RCG) surveys carried out periodically with a sample of recent graduates, most of whom prepared for teaching. A promising future data source will be the Baccalaureate and Beyond Longitudinal Study, which is scheduled to replace the RCG in 1994.

*• *New hires from the reserve pool can be tracked backward* to study their career histories prior to entering or reentering teaching. The SASS teacher survey instrument will provide data on age of entry or reentry, time spent away from teaching, what new hires were doing before they took teaching positions, and subject areas taught. Components of the reserve pool that can be covered in this way include both reentrants and people who were certified but had not taught. Monitored over time, these data will begin to shed light on the extent to which the reserve pool is adequate or exhausted for certain subject areas or geographic areas.

*• *Track persons certified by a given state who are not currently teaching in that state.* Such persons constitute an important component of the reserve pool at the state level. Using data from state certification files, some states can track certificants who still live in the state and can characterize

that segment of the reserve pool by age, subject specialty, and years of past teaching experience. A survey could determine their interest in teaching or incentives that would encourage them to teach. Studies of teacher supply and demand in Massachusetts (Massachusetts Institute for Social and Economic Research, 1987) and Connecticut (Connecticut State Department of Education, 1985b) illustrate the use of certification data to estimate the size of this component of the reserve pool.

5. *Attrition rates and incentives to leave teaching*

*• *School data on attrition rates.* Data from schools on the distribution of teachers by age, race/ethnicity, sex, and disciplinary area, as well as attrition levels within these categories. Attrition should be classified by retirement or other cause.

The best prospect for obtaining some of these data is probably SASS, which included an attrition-by-field question in the base year survey; the data, however, are of poor quality. The panel urges NCES to simplify the SASS matrix questions on attrition to improve response and to collect these data on a continuing basis. The school questionnaire should be able to separate attrition due to moving to another school or district from leaving the teaching profession completely and reducing the national supply of teachers.

*• *Incentives to leave teaching.* Overall changes in supply are affected by factors that make teaching more or less attractive compared with other occupations. The periodic SASS follow-up surveys of former teachers should provide data on the reasons for attrition on a national scale and increase understanding of the behavioral components of teacher attrition and mobility. SASS should also collect information on salary scales, which could be analyzed in conjunction with salaries in other occupations to learn more about the competitiveness of teachers' salaries to opportunity cost salaries.

• *Information from schools on separation rates of teachers by field of study.* These data, which are being collected by SASS, are needed to understand the effect of separations on the teaching force for different fields of study.

• *Information from states on teacher retirement policies.* Such data would be helpful for use in research on the relation among attrition rates, portability of teacher pensions, and retirement policies. (SASS asks districts about the minimum age, years of service, and penalty for early retirement associated with their retirement plans.) However, knowing state retirement policies does not answer the question fully, since teachers are often covered by various combinations of state, district, and union retirement plans.

Data Related to Quality

The notion of "enough" science and mathematics teachers must be understood in qualitative terms. Therefore, it is imperative to gather data that indicate aspects of quality. Little information exists that helps to define or measure quality at present. Although we do not know what constitutes good teaching, one of the issues is to find out what goes into good teaching. We attempted in Chapter 5 to sort out major ingredients of teacher and teaching quality that call for further data. Better measurement of these ingredients may help identify measures of quality. One group of such components includes school system hiring policies and practices and school-level conditions that can affect quality. A second group of aspects involves the qualifications of newly hired teachers. *To provide indicators of aspects of the quality of teachers and their work environments, the panel recommends that the data listed below be collected and monitored over time.*

1. *School system factors that affect quality.* It is critical to build a foundation of data about school and district practices that have effects on quality. The panel recommends obtaining the following data, in order of priority:

*• *Hiring practices, including timing of offers, and constraints such as internal transfer rules.* SASS does not provide data related to these areas; information may be better obtained through in-depth discussions with a sample of districts (as recommended at the end of this chapter).

*• *Data describing inservice education, laboratory materials, and collegial and administrative support for teachers in place.* School principals and school district officials are probably the best source. The one-time administrative and teacher survey of the High School and Beyond Survey conducted in 1984 was designed to provide such data at the national level. SASS and the NSF surveys of science and mathematics education could be the vehicles for data collection, with important design input from in-depth conferences with districts. The presence of support systems allows teachers to be better teachers and can attract new entrants to the profession; however, they are costly. There may be a trade-off between increasing salary and increasing nonsalary support, for example.

*• *School practices related to time use, class size, teaching load, level of autonomy, opportunities for collaboration and decision making, salary, and other monetary incentives.* Information on such school practices may be relatively easy to obtain from teachers or principals through existing survey programs, particularly the SASS school administrator and teacher questionnaires. Other sources include the teacher questionnaires for NAEP science and mathematics assessments, NELS:88, the High School and Beyond administrative and teacher survey, and the SASS follow-up survey of teachers who remain in teaching.

*• *Teacher assignment or misassignment, by subject, including incidence of out-of-field teaching and use of temporary or emergency certification*. (These data were previously recommended to understand supply.) This information, available from the SASS teacher questionnaire, together with the survey's information on filling difficult vacancies, state certification data, and in-depth discussions with school districts as recommended presently, can be used to measure the prevalence of these types of assignments.

• *Identification by school districts of the major criteria used in teacher selection, by subject, and the weight given to each criterion.* In response to a surplus, districts may change the weighting of their criteria, putting more emphasis on formal credentials and depth of course background, in order to make screening easier. However, this kind of information is likely to be difficult to obtain; it may vary greatly depending on the subject or students to be taught. Data collection should be initiated as a research activity rather than by initiating a statistical time series. SASS asks a general question on the use of certain baseline criteria, but collects no data as recommended here.

2. *Qualifications of teachers.* There are a number of ways to measure and assess teacher qualifications. Some are objective and can be counted; some are subjective and not easily quantified. Some are easily quantified but useful only as a baseline for minimum qualification (such as certification); some require more examination but would also be more informative (such as transcripts). Some indicators can be based on existing standards (such as those of the NSTA), and some on proposed standards (such as those of the Holmes group).

While the recommendations presented here are clearly difficult to implement, a beginning must be made. Data should be collected to indicate the presence and strength of teacher qualifications so that more sophisticated studies to measure the effect of teachers' knowledge can be carried out. Thus, the following kinds of data are recommended in order of priority.

*• *Certification data as an indication of a minimum or baseline level of qualifications.* We note that NCES has implemented the following recommendation from the panel's interim report and urge them to continue this practice:

> *"We recommend that the Center for Education Statistics surveys of teachers regularly include measures of certification (type and subject fields) and that the Center obtain and disseminate available information on state certification policies and practices."*

However, certification data should be distinguished not only by type of certification, as the SASS teacher questionnaire now does, but also by whether certification was earned through an alternative certification program. This data recommendation pertains to the supply as well as the quality aspect of precollege science and mathematics teaching.

*• *Individual transcript data on general intellectual ability and on courses taken in preparation for science or mathematics teaching* would provide the most complete data on teachers' formal qualifications. The panel recognizes the cost and burdens of transcript studies, but considers that such studies for samples of teachers would be valuable at the national level and to individual states. For example, transcripts could be used to study trends in educational backgrounds of new teachers or to compare new teachers and continuing teachers.

*• *Trends in guidelines for prospective teachers in terms of content or course work recommended by science and mathematics professional associations and the extent to which guidelines are used.*

• *Trend data on course requirements for teachers by subject specialty for certification, by level of certificate* (as compared with the requirements recommended by boards and professional groups).

Course requirements for certification indicate the minimum background a teacher must possess, unless the teacher is teaching out of field. To the extent that the actual background of a teacher exceeds that level, quality can be said to be higher. However, course background may be only tenuously related to the effectiveness of a teacher; this is a research area noted below. NASDTEC publishes data periodically on course requirements and tests required for certification in each state.

• *If the Carnegie or Holmes recommendations of subject matter degree were adopted, their implementation should be monitored, together with any changes in the supply or quality of the teaching force* as a result of more rigorous requirements.

General Data Recommendations

In addition to the specific data recommendations above, certain general practices should be followed. The panel recommends gathering and maintaining data, at regular intervals, to indicate trends in demand, supply, and quality, and to use in research. During the period of this study, important initiatives were undertaken by NCES to establish new surveys that better describe teacher supply and demand in the United States. Standing out among these efforts are the Schools and Staffing Survey (SASS), the longitudinal study of college graduates that is scheduled to replace the Recent College Graduates (RCG) survey, and the National Educational Longitudinal Study (NELS:88). (These surveys and other ongoing data sets are described in Appendix B.) The new surveys hold particular promise for identifying a number of key aspects of demand and supply, and they should be carefully maintained and built upon accordingly.

The panel recommends adoption of the following general guidelines for any data collection effort relevant to teacher supply, demand, or quality.

*1. *Emphasize the repeated collection of data over time, in contrast to a one-time effort, in order to permit measurement of changes in supply and demand over time.* Data collection activities now being established, such as SASS and NELS-88, need to be repeated at regular intervals. It is the change in the characteristics of the teachers who apply to or are hired by a school district that reflects changes in supply and demand. If a district is able to hire teachers with a master's degree in mathematics, for example, when in the past it was only able to hire teachers with a bachelor's degree, this indicates an improvement in the supply of teachers to the district, whereas reporting the percentage of teachers who hold master's degrees in mathematics does not indicate much by itself. A simple description of the characteristics of new teachers is insufficient, since those characteristics arise from a dynamic process.

*2. *Disseminate collected data into the public domain in a timely manner and in an easily accessible format.* The easier it is to access data and the more opportunity researchers have to analyze data, the more likely is the discovery of timely policies that may have a positive effect on any supply and demand situation.

Ease of access means not only the announcement that data are available, but also good documentation for the data, and the creation of data dictionaries and computer interfaces to facilitate the use of data. Data on teacher supply, demand, and quality can be made more easily available to researchers and students by routinely providing subsets of data bases, or tapes or disks of sample or complete data sets from surveys, for use on personal computers.

*3. *Focus on subareas of subject matter (e.g., chemistry, physics or calculus, rather than mathematics/science in general) in order to permit specific identification and targeting of areas of shortage or surplus.* Especially at the high school level, teachers of biology, chemistry, physics, or general science are not interchangeable. Thus, data collection that aggregates science teachers or science and mathematics teachers under a single heading is likely to mask shortages of teachers in specialty areas.

4. *Focus on trends for minorities and women—both students and teachers—in the various subject areas at issue.* Minorities compose an increasing share of precollege enrollments. Women take fewer courses in science and mathematics at higher levels than do men. Trends in the participation of these groups in science and math are an important indicator of their future supply in high-technology employment. This supply may also be increased by the availability of teachers of the same ethnicity or sex as role models.

5. *Ascertain, for each data collection activity considered, whether the federal government or another entity could best collect the data.* Most expenditures for precollege education and all personnel decisions are made at the state and local levels. Most detailed data are collected at those levels as well. Problems arise, however, when such data are aggregated to obtain a national perspective of the conditions of teacher supply and demand. School districts, for example, are not likely to record whether a new hire is new to teaching or simply new to the district. National data on changes in supply, however, should exclude teachers who simply move from district to district, since they are not an element in measuring the national supply of teachers. Data on the age and experience of employed teachers, however, exist at the state and district levels and should not be recollected to obtain national measures.

RESEARCH ISSUES IDENTIFIED BY THE PANEL

A number of important issues affecting supply and demand for science and mathematics teachers are not well understood and are beyond the scope of existing data and models. These are research questions that need to be understood in order to determine the types of data needed to properly model demand, supply, or quality. They are likely to require long-term research. The panel also touches on their relation to student outcomes. The panel did not attempt to develop a comprehensive list of research issues, but in the course of panel discussions a variety of research topics was noted and are reported here.

Resources for Research

For the present, it is the panel's conviction that the research base is inadequate to support the development of behavioral models of teacher supply and demand. We have identified a range of issues about how teacher labor markets work and how they affect teacher supply, demand, and quality. Further research on these issues is needed to enable the development of causal models of teacher supply and demand.

**The panel recommends that the National Science Foundation stimulate research on behavioral models of teacher supply and demand, and increase the amount of support for such research.*

The research issues pertaining to each of the topics—demand, supply, and quality—are presented below, with asterisks marking those we consider highest in priority. We also note the importance of research related to student outcomes.

Demand

For development of improved models for longer-term projections, research is needed on the behavioral factors that influence the demand for teachers, particularly teachers of science and mathematics, in the higher grades, and on the pupil-teacher ratio as both a dependent and independent factor in creation of demand. The panel therefore lists the research topics pertinent to teacher demand that were identified in the panel's interim report, in order of priority:

*1. *The behavioral determinants of student selection of science and mathematics courses* at the secondary school level, including the effects of changes in graduation requirements and of student preferences for subject areas;

*2. *The behavioral determinants of parental and student preferences for private and public schooling*;

*3. *The determinants of pupil-teacher ratios*, especially the adjustment lags in those ratios as enrollments change and/or the teaching force changes in demographic composition;

*4. *The impact on high school dropout rates* of such factors as changes in graduation requirements, labor market conditions, and the demographic composition and family circumstances of the school-age population; and

*5. *The relationship of changes in demand for courses to changes in pupil-teacher ratios* and the resulting derived demand for full-time-equivalent teachers of mathematics and science at the secondary school level.

In amplification of issue (3) above, it is noted that other factors can affect pupil-teacher ratios: changes in the school budget and changes in staffing patterns, class size, and teaching loads. A closer analysis of factors other than changing enrollments that influence pupil-teacher ratios is an area for further research. In periods of teacher or budget shortages, the ratio (or class size) can be increased. When demand slackens, teachers might stay, and so it would drop. Research on these coping responses and pupil-teacher ratio changes could lead to more accurate assumptions about the pupil-teacher ratio for demand models. Of particular interest would be a disaggregation by subject, so one could focus on pupil-teacher ratios for science and mathematics.

Supply

Although incentives or disincentives exist in schools as they do in most organizations, we lack detailed knowledge of how they affect supply. A variety of behavioral and environmental factors influence the number of individuals willing to teach: compensation (both salary and benefits), working conditions, availability of other jobs in the labor market area, cost of educational training, and state and local policies for educational professional personnel. These behavioral factors affect not only the actual supply of teachers, but also the retention of current teachers. These behavioral factors are particularly important because many of them are levers that policy makers can use to change the supply of teachers. Much can be learned from in-depth conferences with school district officials, which we recommend at the end of this chapter. Recommended studies of the effect of these behavioral factors are listed below in order of priority.

*1. *Incentives that affect individual decisions to enter teaching, to leave teaching and move to a different occupation, or to retire.* For the first, what are the effects of salary, working conditions, location, alternative nonteaching opportunities, etc., on the decision to accept an offer to teach and on the overall acceptance rates? For the second, what are the effects of salary, instructional support, working conditions, alternative opportunities, etc., on whether, and when in the career cycle, one decides to leave and on the overall retention rates of teachers? For the third, research is needed on the relationship between state separation rates for retirement and individual reasons for retiring, external shock variables, and incentives and disincentives for retention and retirement. Related issues are the effect of separation rates on individual school districts and on the teaching force for different fields of study.

*2. *Supply potential of the reserve pool and supply.* Because the reserve pool is one of the two major sources of teachers, and because the other source—new certificants—is decreasing in number, research to assess the supply potential of the reserve pool is of increasing importance. Studies of the reserve pool might include the effects of incentives, such as salary increases, on attracting individuals from the reserve pool. To determine whether entry rates to teaching from the reserve pool are influenced by salary increases, districts that have had large salary increases (e.g., Rochester) could be studied. Another example: What are the effects of limited mobility of teachers in the reserve pool on the supply potential of the reserve pool? It appears that in general individuals are willing to move only a small distance to accept a teaching position. A study of the labor markets for urban areas, suburban areas, small towns, and rural areas could help determine what that distance is. These examples are only a few among numerous important aspects of the reserve pool in need of research.

*3. *School districts experiencing supply/demand problems.* The information collected in SASS can be used to identify such school districts for special studies. The supply and demand situation for science and mathematics teachers is likely to be quite different in different geographic or labor market regions (e.g., inner city, rural, or high-income suburban).

Examination of subsamples of districts, including in-depth inquiries, can produce valuable special reports if done on a timely basis. They may provide information for policy use in ameliorating the problems, and they can also help determine appropriate categories for disaggregation of data in publications. The NCES district conferences we recommend later in this chapter can be designed to coordinate with the study of particular groups of districts.

4. *Alternative career decisions that minority college students and graduate students are making.* Why are fewer such students obtaining degrees to teach? Why are they choosing other positions rather than teaching jobs? Why are they leaving teaching? Research should probe into the first two questions to identify the alternatives minorities choose and the perceived opportunity costs that draw them away from teaching science or mathematics. Such knowledge can help form strategies to attract minorities into teaching and, indirectly, to increase the overall supply pool. The last question could be addressed with data from the SASS questionnaires. The best data source for the earlier question is the set of follow-up surveys of the NLS-72.

5. *Effect of increasing certification requirements on the incentive to obtain a teaching certificate or to apply for a teaching position.*

6. *Incentives that attracted the recipients of NSF's Presidential Science Awards to teaching and those that keep them in the teaching profession.* The professional development activities of the recipients of mathematics awards have been studied (Yamashita, 1987).

Quality

Neither teacher quality nor teaching quality is a term that lends itself readily to precise definition. Teacher quality refers to the knowledge, skill, and general ability level of the teacher. We believe that measurement of teacher quality is important because some of the measures of teacher quality seem to be important factors in determining who goes into teaching and who finds better opportunities elsewhere. Thus teacher quality is an important variable in teacher supply models, and we need to understand the responsiveness of teacher quality to incentives. Obviously, teacher quality is determined in part by the quality of teacher preparation programs; the issue is the degree to which these programs prepare teachers to be effective pedagogues in transmitting knowledge about mathematics or science.

Teaching quality is also of direct concern to the panel, since it is a direct measure of the degree to which a teacher of mathematics or science is able in the school setting to lead students to a better understanding of mathematics or science. Teaching quality, as we see it, derives from several sources and can be measured by different types of data pertaining to the school setting in which classroom teaching takes place. Teaching quality is affected by school and district policies and practices, such as selection of teaching materials, allocation of time to various subjects, availability of laboratory facilities, and the teacher's degree of autonomy in the classroom. It is also affected by curricular structure and processes and by teacher characteristics such as subject matter competence and ability to facilitate learning, which in turn are affected by the quality of teacher training.

A frequently used measure of teacher quality is teacher qualifications–courses taken, credentials received, etc. While we recognize that there is some link between teacher qualifications and both teacher quality and teaching quality, it is the panel's view that the linkage is apt to be loose rather than tight; again, that is clearly a topic for research.

In the course of panel discussions on these issues, we noted several studies related to teacher quality or teaching quality that could be pursued, and we list them in order of priority.

*1. *Study the effectiveness of practices that schools and school districts have employed to improve quality*–for example, the mentor schools in education in Dade County, Florida.

*2. *Examine the inservice training practices for science and mathematics teachers* that are provided by elementary and secondary schools, to identify programs that seem to be effective and to understand reasons why some programs appear to work while others do not.

*3. *Study teachers' transcript records* to determine the degree to which transcripts can be used as an accurate reflection of subject matter knowledge.

*4. *Study the methodological curriculum in teacher training institutions* to assess the degree to which these institutions vary in their emphasis on pedagogical theory compared with pedagogical practice.

5. *Compare the academic backgrounds of teachers who leave teaching and those who stay.* (Substudies based on teacher transcripts could be conducted for teachers identified in the Schools and Staffing Survey.)

6. *Conduct a follow-up at the schools of the recipients of the Presidential Awards in Science and Mathematics Teaching* to gain insights into factors that might improve quality at the school level by noting how the award money was actually spent. The supply-related research issue suggested above could be coordinated with this in a single research project focusing on the award recipients.

7. *Measure the extent of movement of teachers* within school systems from elementary to middle to high school teaching and assess whether these transitions are eroding the average level of content background for secondary science and mathematics teachers and for the body of remaining elementary teachers who do not transfer.

Student Outcomes

Although the focus of the panel's study is on the supply, demand, and quality of teachers and on the data needed to monitor these phenomena, it is clear to the panel that their ultimate usefulness lies in the effect of these characteristics on learning. Thus, it is important to measure supply, demand, and quality because it is presumed that these factors are linked to student learning outcomes. But that linkage needs to be explicit; it constitutes an important area in which substantial research efforts need to be carried on. Some aspects of research that would attempt to relate teacher characteristics, school environment variables, and home environment variables to student outcomes and to recognize the importance of variation in school outcomes are described below.

1. *Teacher characteristics.* Research to date has not shown a clear link between teacher characteristics and student outcomes. As indicated in Chapter 5, it appears that verbal ability, the number of mathematics or science credits, recent educational experience, professional involvement, years of teaching, and attitudes toward teaching may exhibit some positive relationship, often weak, to student performance. A better understanding of the relationship between teacher characteristics and student outcomes is needed. As a start toward research on this issue, the National Assessment of Educational Progress (NAEP) teacher questionnaire for 7th- and 11th-grade science and mathematics teachers, from the latest science and mathematics assessment (in 1985-86) can be studied, in conjunction with their students' science and mathematics NAEP test scores. The next NAEP science and mathematics assessment is planned for 1990. In addition, the National Education Longitudinal Study of 1988 (NELS:88) includes measures of student outcomes together with a set of teacher characteristics and characteristics of schools and districts. For the research to be more meaningful, measures should be obtained through records such as transcripts rather than through survey questions.

2. *School environment factors.* Research should also be concerned with school environment factors that can affect student outcomes, for example, the constraints placed on teachers by school or district practices such as the degree of mentoring provided to new teachers, teachers' opportunities to interact with other teachers, and the allocation of classroom hours among mathematics, science, and other subjects.

3. *Home environment variables.* Learning is influenced not only by teacher and teaching characteristics, but also–and perhaps primarily–by the characteristics of the student's home environment. Before learning takes place in formal school settings, it takes place in the home, and home environments continue to play a role in student learning throughout the entire developmental process. Thus, incorporating the influence of home

environment variables (e.g., time spent by parents with children and the beliefs and expectations of parents for their children) in studies of student outcomes is crucial to understanding the true influence of teachers and teaching factors on learning.

4. *Variation in school outcomes.* The objective of school is to facilitate learning and, from the perspective of the panel, to promote learning in science and mathematics. But from that perspective, facilitation can mean either improving the average outcome or reducing the variation in outcomes. These objectives can sometimes conflict, and part of what most people mean by effective teaching is to find ways to increase the minimum that all students master while not restricting the opportunities of the more able or the more rapid learners. Much existing research has focused largely on the influence of schools on average outcomes, and has not recognized the importance of variation in outcomes.

**The panel recommends that further research be conducted on the relationship of measurable characteristics of teachers of mathematics and science and home and school environmental factors to educational outcomes of students in these fields. This research should explore variation in outcomes as well as average outcomes.*

Although we recognize the difficulty of conducting controlled experiments in education, we believe such experiments could be particularly useful in studying the relationship between measurable teacher qualifications and student outcomes.

RESEARCH FACILITATION

Educational research is carried out by a number of constituencies: federal, state and local government, research organizations under contract to the government, and academic institutions. Research carried out by governments or research organizations is normally designed to answer specific questions. For example, a research organization is given a grant or contract to study the usefulness of indicators in education.

Graduate Student Research

Research at academic institutions is carried out in the form of doctoral dissertations or by faculty interested in particular aspects that relate to a specific area of knowledge. For example, a doctoral candidate whose field of interest is gender may be interested in determining gender differences with respect to mathematics background for teachers. The number of doctoral dissertations is large and reflects a rich source of highly trained individuals.

In order to attract this group to work on problems of educational interest, student support should be available.

**The panel recommends that the Office of Educational Research and Improvement within the Department of Education create a program of doctoral graduate student support (training grants) in education statistics.*

The training grant program in the health sciences (biostatistics) has been very successful in attracting to the field of biostatistics a large number of individuals, many of whom are currently employed by the National Institutes of Health, other health organizations, and the pharmaceutical industry. This has changed the level of sophistication in these fields and permitted studies to answer questions on the health status of our society. A comparable program in education statistics could bring to education a parallel group of talented researchers.

Data Bases for Personal Computers

To carry out a doctoral dissertation related to teacher quality or models of teacher supply and demand, access to relevant data is needed.

The panel recommends that data from education agencies and studies relating to education be made available in the form of tapes of the complete data sets, as well as user-friendly disks of data samples.

It is particularly important that the documentation be understandable to the researcher without too great an investment of time. An analysis of a large data set can be a very time-intensive process, often taking a year to complete. Such a required time expenditure would not permit graduate students sufficiently ready access for use on a dissertation.

Small data sets should be made available for classroom and textbook use. This would have the effect of making education data more visible to a wider audience and ensuring more extensive analysis of the information.

State Data Bases

The national data bases we have described, as valuable as they will be, will not detail state labor markets or labor markets by field of study. Studies by Murnane and his colleagues (1988, 1989) show that when data are desired for longer time periods or for variables beyond those collected in these national studies, or when greater disaggregation is needed, state data bases become the most useful existing resource. For this reason, the panel surveyed the state education agencies (SEAs) concerning their available data files on public school professional personnel. The results of this survey are summarized in Appendix C. The appendix shows the earliest date for which data are available and the availability of selected data items. In Massachusetts, the Massachusetts Institute for Social and Economic Research (MISER) used files of the state certification board and the state retirement system to study the supply and demand for teachers in the state of Massachusetts. This study is being expanded to all states in the New England region.

There are indications in state studies by Murnane and his colleagues and in Heyns' analysis of the NLS data (1988) that there are large differences between elementary and secondary teachers in career paths, decisions to stay in teaching, and patterns of reentry to teaching. The studies by Murnane and his colleagues also indicate that both states and fields of study show differences along these variables. The studies are also useful in showing how to combine single-year record tapes of state agencies into multiyear career history files, making it possible to study the attrition and retention of teachers.

Limitations of Univariate Indicators

The panel sounds a note of caution about interpretation of univariate indicators of quality and about analyses that fail to consider the context in which quality is measured. The quality of science and mathematics instruction is multifaceted, and single variables are best studied in a multifaceted context.

An indicator of quality is offered as an example. Membership in mathematics or science teacher professional associations, attendance at workshops sponsored by these associations, and payments by school districts for advanced training in teachers' specialties during the summer are univariate indicators that appear amenable to data collection. Although it would seem that these are measures of quality, that may not always be the case. If these data are cross-classified by district characteristics, they might indicate differences. Whether or not these differences are differences in quality is a moot point. For instance, districts with small enrollment might support advanced training in the summer to compensate for the isolation of their teachers. Districts with high socioeconomic status might do so to reward their outstanding teachers. Still other school districts finding it difficult to recruit teachers might offer such training as an incentive for hiring, as is done by the Dade County and New York City school districts.

This leads to a caution about interpretation of univariate indicators of quality and about policy analysis that fails to account for their value by considering the context in which quality is measured. Frequently cited measures of science education quality are the number of physics courses offered by a high school and the number of teachers certified to teach physics. Alternatively, frequently cited measures of low quality are the number of physics courses taught by an out-of-field teacher certified in another physical science or the absence of physics courses. Under this definition, a large proportion of the high schools in this country would be rated as of low quality in science education, and a large proportion of the nation's students attend these schools.

Before making a policy prescription it is important to classify districts by characteristics that may affect the behavior observed. The low score on this measure is attributable to the low enrollment size of these schools, which is too small to support a full-time physics teacher. In this example, size was an intervening variable, which should lead us to expect split teaching assignments and few courses in physics in small high schools and to consider policy initiatives appropriate to small schools. Obviously, the explanation for the low score on this quality measure—small schools—does not change the low score nor does it change the quality of the part-time teachers.

As another example of the caution needed in using univariate indicators, high achievement test scores have often been associated with schools in high socioeconomic neighborhoods. Analysts might find a high correlation between some educational practice or teacher background variable and student achievement. The teacher background variable might relate to the practice. The high socioeconomic neighborhood attracts teachers with the background variable. A multiple regression would indicate that all three variables—socioeconomic neighborhood, teacher background, and educational practice—relate to high achievement test scores. If only one of these variables is studied, the explained variance will be overstated. Quality instruction in science and mathematics is multifaceted, and single variables are best studied in a multifaceted context.

FACILITATION OF INFORMATION EXCHANGE AMONG DISTRICTS, STATES, AND THE NCES

As evident throughout the report, the activities the panel carried out to obtain information about the flow of teachers and the quality of teachers in individual school districts influenced our thinking in many ways. The 16,000 school districts in this country operate relatively independently. The staffing problems they encounter vary widely, and the actions taken by district superintendents and personnel directors to address these problems are both innovative and varied. Both applicants and school systems have effective means of coping with the uncertainty of budgets and contracts and adjusting to institutional barriers (e.g., use of the substitute pool to stockpile place-bound potential teachers, use of graduate students to teach part time, cooperative arrangements with local industry). Much of the information about school district actions to address staffing problems will not be captured by the SASS. A few illustrations follow.

- New York City has hundreds of science and mathematics teachers who are teaching out of field, but not because of a teacher shortage. The school principals have not requested replacements for these teachers because they are effective in working with students in inner-city schools, whereas the ability of certified replacements to control the classroom is an unknown. Yet incidence of out-of-field teaching is one of the measures of shortage in use.
- The personnel officer of a middle-sized Texas district asserts firmly that the district has no shortage of science and mathematics teachers, and yet they rely heavily on an aggressive national recruiting program.
- The panel selected a pair of adjacent districts in Maryland for in-depth case studies because it was thought they would draw on the same labor market. In fact the large urban district recruited at numerous job fairs (occasions at which as many as 20 teacher-training institutions gather their graduates on one campus for a one- or two-day meeting) in areas as far away as Illinois, Michigan, northern New York, and North Carolina. The smaller, semirural district recruited in rural areas of the state and neighboring states. The more rural district was looking for teachers who would be content to live in a rural area and whose values would be similar to those held by the community.
- The nature and timing of collective bargaining increases the difficulty of making accurate demand projections. Since many contracts are negotiated in late spring or during the summer, and since clauses typically offer benefits like improved health care provisions to any teachers under contract on the date the agreement is signed, teachers who intend to resign wait until the contract is completed before giving notice. This leads to an underestimate of attrition for demand projections.
- Hiring is a year-round process for the seven large districts that participated in the panel's conference, and "demand" is not a static number that measures need only in the fall of each year. In fact these districts do only about half of their hiring for September.
- The Dade County School System found that widespread recruitment was not as successful as anticipated. Although they recruited successfully in the northern tier of states, retention was a problem because the new recruits could not cope with the multicultural student body, the heat, the lack of seasons, and homesickness.

The case studies and the conference of personnel directors vividly demonstrated to the panel the diversity of practices and styles and the diversity of labor market situations that characterize the nation's school districts. The panel believes that NCES could profit from frequent interactions with school district personnel and could play a valuable role as a broker between data producers and data users in the states. A useful mechanism for such interaction would be conferences of district and/or state personnel. At least three types of conferences are envisioned:

1. An annual conference structured to help NCES in design and analysis of SASS. Attendees at this conference would be a mix of district superintendents and personnel directors from districts in the SASS sample. The group should be small enough (7-10 individuals) to permit roundtable discussion. The district personnel would be asked to discuss what is going on in the district with respect to teacher supply, demand, and quality that is not revealed by the data on the SASS forms. This information could be used to provide a framework for analysis of SASS data and caveats to accompany the analysis, and possibly to identify items that should be added to or deleted from the SASS forms.

2. A conference designed to facilitate analysis of teacher supply and demand at the state and district levels. Attendees would be state personnel who prepare supply-and-demand models and individuals experienced with modeling who would be sensitive to implicit assumptions in the models that might not be appropriate for use in some states. An exchange of ideas among these individuals could lead to improvements in state models and in models that states prepare for their individual districts. (When the panel convened the conference of personnel directors from large school districts, we were surprised to learn that these individuals had never met, yet they had problems that were unique to big districts in both nature and magnitude. One of the major benefits of the meeting to them was the opportunity to share problems and solutions. The conference was of sufficient value to them that they have instituted an annual meeting involving a larger number of districts.)

3. A conference designed to stimulate communication between state data producers and district data users. The conference of personnel directors of large school districts provided a striking example of the potential benefit of such conferences. The personnel director from New York City suggested that it would be helpful to have some central organization collect and disseminate information on the number of persons enrolled in teacher training programs, by institution, as contrasted with the currently available data on education majors. One of the panel members knew that the desired information is currently collected by the state education agency and arranged to send it to the district.

Brief reports of the conferences should be prepared so that districts and states that were not represented among the conferees could also benefit from the exchange of ideas.

By maintaining frequent contact with the individuals who make the decisions that affect teachers, much can be learned about the flow of teachers through the school system, and the quality adjustments made in this flow by aggressive recruiting, raising or lowering standards in hiring, providing inservice training, and using incentives to attract and retain teachers.

**The panel recommends that the National Center for Education Statistics convene (a) an annual conference of district personnel who are responsible for the decisions that affect teacher supply, demand, and quality to maintain an awareness of the current issues in teacher supply and demand; (b) periodic conferences of state personnel who prepare state and local supply and demand projections to facilitate improvement in these models; and (c) occasional conferences to promote communication between state personnel who produce relevant data and district personnel who would find these data useful in their recruitment activities and in development of district policies concerning teachers.*

The panel learned much from the interaction with district personnel. We believe that staff of the National Center for Education Statistics would find it equally rewarding and that the center's surveys and analyses would be enriched by such interaction.

References

This list includes documents referenced in the report as well as documents known to the panel that pertain to the supply and demand for precollege teachers.

Akin, James N.

1983a Teacher supply/demand by field and region. *Education Week* Feb. 16:16-17.

1983b *Teacher Supply/Demand 1983.* Madison, Wisconsin: Association for School, College and University Staffing, Inc.

1984 *Teacher Supply/Demand 1984—A Report Based Upon an Opinion Survey of Teacher Placement Officers.* Madison, Wisconsin: Association for School, College and University Staffing, Inc.

1985 *Teacher Supply/Demand 1985—A Report Based Upon An Opinion Survey of Teacher Placement Officers.* Madison, Wisconsin: Association for School, College and University Staffing, Inc.

1986 *Teacher Supply/Demand 1986—A Report Based Upon An Opinion Survey of Teacher Placement Officers.* Madison, Wisconsin: Association for School, College and University Staffing, Inc.

Alabama State Department of Education

1982 *Comparisons of Supply and Demand.* Montgomery, Alabama: Author.

Alexander, Lamar, and H. Thomas James

1987 *The Nation's Report Card—Improving the Assessment of Student Achievement.* Washington, D.C.: U.S. Department of Education.

American Association of Colleges for Teacher Education (AACTE)

1984 *Teacher Education Policy in the States: 50-State Survey of Legislative and Administrative Actions.* Washington, D.C.: Author.

1987a *Minority Teacher Recruitment and Retention: A Call for Action.* Policy Statement, September. Washington, D.C.: American Association of Colleges for Teacher Education.

1987b *Teaching Teachers: Facts and Figures.* Washington, D.C.: American Association of Colleges for Teacher Education.

1989 Data in the form of printed figures and tables. July. American Association of Colleges for Teacher Education, Washington, D.C.

American Council on Education

1988 Study cites growing shortage of minority teachers. *Higher Education and National Affairs* 37(19)(October 31):3,8.

American Institute of Physics

1988 *Physics in the High Schools. 1986-87 Nationwide Survey of Secondary School Teachers of Physics*. M. Neuchatz and M. Covalt, authors. Pub. No. R-340. November. New York: American Institute of Physics.

Antos, J., and S. Rosen

1975 Discrimination in the market for public school teachers. *Journal of Econometrics* 3(2):123-150.

Arrow, Kenneth J., and William M. Capron

1959 Dynamic shortage and price rises: the engineer-scientist case. *Quarterly Journal of Economics*.

Ascher, William

1979 Problems of forecasting and technology assessment. *Technological Forecasting and Social Change* 13:149-156.

Association of Teacher Educators

1987 *Teacher Induction—A New Beginning*. Reston, Virginia: Association of Teacher Educators.

Atkin, J. Myron

1981 Who will teach in high school? *Daedalus* 110:91-103.

Baratz, Joan C.

1986 *Black Participation in the Teacher Pool*. January. Paper prepared for the Carnegie Forum's Task Force on Teaching as a Profession. Rochester, New York: Carnegie Forum on Education and the Economy.

Barro, Stephen M.

1986 *The State of the Art in Projecting Teacher Supply and Demand*. Paper prepared for the Panel on Statistics on Supply and Demand for Precollege Science and Mathematics Teachers, Committee on National Statistics. Washington, D.C.: SMB Economic Research, Inc.

Barro, Stephen M., and Stephen J. Carroll

1975 *Budget Allocation by School Districts: An Analysis of Spending for Teachers and Other Resources*. R-1797-NIE. Santa Monica, California.: The RAND Corporation.

Baugh, William H., and Joe A. Stone

1982 Mobility and wage equilibrium in the educator labor market. *Economics of Education Review* 2(3):253-274.

Benderson, Albert

1982 *Teacher Competence*. *Focus* (10). Princeton, New Jersey: Educational Testing Service.

Berliner, David C.

1986 In pursuit of the expert pedagogue. *Educational Researcher* August/September, Pp. 5-13.

Berry, Barnett

1984 *A Case Study of the Teacher Labor Market in the Southeast*. Occasional Papers in Educational Policy Analysis No. 413. Research Triangle Park, North Carolina: Southeastern Regional Council for Educational Improvement.

1985 *Understanding Teacher Supply and Demand in the Southeast: A Synthesis of Qualitative Research to Aid Effective Policymaking*. Research Triangle Park, North Carolina: Southeastern Regional Council for Educational Improvement.

Berryman, Sue E.
1983 *Who Will Do Science?* New York: The Rockefeller Foundation.

Bird, Ronald E.
1984 *Report and Evaluation of Current Information Regarding Teacher Supply and Demand.* Occasional Papers in Educational Policy Analysis, No. 8. Research Triangle Park, North Carolina: Southeastern Regional Council for Educational Improvement.

Bishop, A. J.
1982 Implications of research for mathematics teacher education. *Journal of Education for Teaching* 8(2):119-135.

Blank, Rolf K.
1986 *Science and Mathematics Indicators: Conceptual Framework for a State-Based Network.* Washington, D.C.: Council of Chief State School Officers.

Blank, Rolf K., and Senta A. Raizen
1985 Background Paper for a Planning Conference on a Study of Teacher Quality in Science and Mathematics Education. Committee on Research in Mathematics, Science, and Technology Education, National Research Council, Washington, D.C.
1986 Background Paper for a Planning Conference on a Study of Teacher Quality in Science and Mathematics Education. For February 7-8, 1986, meeting of the Panel on Statistics on Supply and Demand for Precollege Science and Mathematics Teachers, National Research Council, Washington, D.C.

Block, Erich
1989 Engineering and national interests. *The Bridge* 19(1)(spring):11-15. Washington, D.C.: National Academy of Engineering.

Bloland, P. A., and T. J. Selby
1980 Factors associated with career choice among secondary school teachers: a review of the literature. *Educational Research Quarterly* 5:13-23.

Boozer, Robert F.
1985 *Supply and Demand—Educational Personnel in Delaware 1984-85.* July. Dover: Delaware Department of Public Instruction.
1986 *Teacher Supply and Demand: Definitions, Data, Models and Instrumentation.* Paper presented to the National Research Council Panel on Statistics on Supply and Demand for Precollege Science and Mathematics Teachers, February 7. Dover, Delaware: Delaware Department of Public Instruction.

Boyer, Ernest L.
1983 *High School: A Report on Secondary Education in America.* The Carnegie Foundation for the Advancement of Teaching. New York: Harper and Row.

Brunner, Ronald D., J. Samuel Fitch, Janet Grassia, Lyn Kathlene, and Kenneth R. Hammond
1987 Improving data utilization: the case-wise alternative. *Policy Sciences.* 20:365-394.

Bureau of the Census
1978 *Estimates of the Population of the United States, by Age, Sex and Race.* CPR, P-25, No. 721. April. Washington, D.C.: U.S. Department of Commerce, Bureau of the Census.
1980 *Major Field of Study of College Students: October 1978.* Current Population Reports. Series P-20, No. 351. May. Washington, D.C.: U.S. Department of Commerce, Bureau of the Census.

1982 *Money Income and Poverty Status of Families and Persons in the United States: 1982.* Current Population Reports, Series P-60, No. 140. Washington, D.C.: U.S. Department of Commerce, Bureau of the Census.

1984a *Projections of the Population of the United States, by Age, Sex and Race: 1983 to 2080.*

1984b *Marital Status and Living Arrangements.* Series P-20, No. 399. March. Washington, D.C.: U.S. Department of Commerce, Bureau of the Census.

1986 *Money Income and Poverty Status of Families and Persons in the United States: 1985.* Series P-60, No. 154. August. Washington, D.C.: U.S. Department of Commerce, Bureau of the Census.

1987 *Estimates of the Population of the United States, by Age, Sex and Race: 1980 to 1986.* Current Population Reports, Series P-25, No. 1000. February. Washington, D.C.: U.S. Department of Commerce, Bureau of the Census.

1988 *The Hispanic Population in the United States.* Current Population Reports, Series P-20, No. 431. August.

1989 *Statistical Abstract of the United States, 1988* December. Washington, D.C.: U.S. Department of Commerce, Bureau of the Census.

Burke, Peter J.

1987 Inherited merry-go-rounds. In *Teacher Development: Introduction, Renewal and Redirection.* New York: Falmer.

California Round Table on Educational Opportunity

1983 *Improving the Attractiveness of the K-12 Teaching Profession in California.* Sacramento: California State Department of Education.

California State Commission on Teacher Credentialing

1986 Minutes and Survey of Participating Agencies, Conference on Teacher Supply and Demand Memorandum from Richard K. Mastain. California State Commission on Teacher Credentialing, Sacramento.

Cagampang, Helen, Walter I. Garms, Todd J. Greenspan, and James W. Guthrie

1986 *Teacher Supply and Demand in California: Is the Reserve Pool a Realistic Source of Supply?* August. Berkeley: Policy Analysis for California Education, University of California.

California State Department of Education

1967 *Teacher Supply and Demand in California, 1965-1975.* Sacramento, California: Author.

1982 *Characteristics of Professional Staff in California Public Schools.* Sacramento, California: Author.

1983 *Supply and Demand for Bilingual Teachers in School Districts in California, An Update Report: 1982-83.* Sacramento, California: Author.

1984 *The Teacher Shortage in California: A Preliminary Analysis.* Prepared by James Fulton. Sacramento, California: Planning, Evaluation, and Research Division, Author.

Capper, Joanne

1987 *A Study of Certified Teacher Availability in the States.* February. Washington, D.C.: Council of Chief State School Officers.

Carey, Neil B., Brian S. Mittman, and Linda Darling-Hammond

1988 *Recruiting Mathematics and Science Teachers Through Nontraditional Programs: A Survey.* May. Santa Monica, California: The RAND Corporation.

Carnegie Forum on Education and the Economy

1986 *A Nation Prepared: The Report of the Task Force on Teaching as a Profession.* New York: Carnegie Forum on Education and the Economy.

Carnegie Task Force on Teaching as a Profession
1986 *A Nation Prepared: Teachers for the 21st Century*. New York: Carnegie Forum on Education and the Economy.

Carroll, C. Dennis
1985 *High School and Beyond Tabulation: Background Characteristics of High School Teachers*. Washington, D.C.: National Center for Education Statistics, U.S. Department of Education.

Carroll, Stephen J.
1973 *Analysis of the Educational Personnel System: III. The Demand for Educational Professionals*. Santa Monica, California: The RAND Corporation.

Carroll, Stephen J., and Kenneth F. Ryder, Jr.
1974 *Analysis of the Educational Personnel System: V. Supply of Elementary and Secondary Teachers*. R-1341-HEW. Santa Monica, California: The RAND Corporation.

Cavin, Edward S.
1986 *A Review of Teacher Supply and Demand Projections by the U.S. Department of Education, Illinois, and New York*. Draft paper submitted to the Panel on Statistics on Supply and Demand for Precollege Science and Mathematics Teachers, Committee on National Statistics. Princeton, New Jersey: Mathematica Policy Research, Inc.

Cavin, E. S., R. J. Murnane, and R. S. Brown
1984 *How Enrollment Declines Affect Per Pupil Expenditure Levels in Public School Districts*. Final report to the National Institute of Education, U.S. Department of Education. Princeton, New Jersey.: Mathematica Policy Research, Inc.
1985 School district responses to enrollment changes: the direction of change matters! *Journal of Education Finance* 10(4):426-440.

Cawelti, G., and J. Adkisson
1985 ASCD study reveals elementary school time allocations for subject areas: other trends noted. *Supplement to ASCD Update*. Alexandria, Virginia: Association for Supervision and Curriculum Development.

Center for Education Statistics
1987a *The Condition of Education—A Statistical Report*. Washington, D.C.: U.S. Department of Education.
1987b *Digest of Education Statistics, 1987*. Washington, D.C.: U.S. Government Printing Office.
1987c *Integrated Postsecondary Education Data System—Glossary*. Washington, D.C.: U.S. Government Printing Office.
1987d *The National Longitudinal Study of the High School Class of 1972 (NLS-72)—Fifth Follow-Up (1986). Data File User's Manual*. December. Washington, D.C.: Center for Education Statistics, U.S. Department of Education.
1987e *Teachers in Elementary and Secondary Education*. Based on *1983-84 Survey of Teacher Demand and Shortage*. Historical Report CS 87-324h. Washington, D.C.: Center for Education Statistics, U.S. Department of Education.
1987f *Trends in Bachelors and Higher Degrees 1975-1985*. Washington, D.C.: U.S. Government Printing Office.
1988a *Conference on Elementary and Secondary Education Statistics Program, January 25-26*. Draft report by S. A. Raizen, March 15. Washington, D.C.: Center for Education Statistics, U.S. Department of Education.

1988b *Racial/Ethnic Data for 1984 Fall Enrollment and Earned Degree Recipients for Academic Year 1984-85.* Tabulation CS 88-200. Washington, D.C.: Center for Education Statistics, U.S. Department of Education.

1988c *Residence of First-Time Students in Postsecondary Education Institutions, Fall 1986.* Tabulation CS 88-224. Washington, D.C.: Center for Education Statistics, U.S. Department of Education.

1988d *Trends in Minority Enrollment in Higher Education, Fall 1976-Fall 1986.* Survey Report CS 88-201. Washington, D.C.: Center for Education Statistics, U.S. Department of Education.

Center for Policy Research in Education (CPRE)

1989 *The Implementation and Effects of High School Graduation Requirements: First Steps Toward Curricular Reform.* Research Report Series RR-011. February. New Brunswick, New Jersey: Rutgers University.

Center for Statistics

1985 *Factors Associated with Decline of Test Scores of High School Seniors, 1972 to 1980.* CS 85-217. Washington, D.C.: Center for Statistics, U.S. Department of Education.

1986a *Digest of Education Statistics, 1985-86.* Washington, D.C.: U.S. Government Printing Office.

1986b *Plan for the Redesign of the Elementary and Secondary Data Collection Program.* Working Paper, March 18. Washington, D.C.: U.S. Department of Education.

1986c *Postsecondary Credits Earned by 1980 High School Seniors Who Received Bachelors' Degrees.* Tabulation of High School and Beyond by J. A. Owings. LSB 86-3-3. Washington, D.C.: Center for Statistics, U.S. Department of Education.

1986d *Public High School Graduation Requirements.* Report No. CS 86-225b. September. Washington, D.C.: U.S. Department of Education.

Center for the Study of the Teaching Profession

1987 *Annual Report, October 1986 to September 1987.* Santa Monica, California: The RAND Corporation.

Champagne, Audrey B., and Leslie E. Hornig, eds.

1986 *This Year in School Science 1985—Science Teaching.* The Report of the 1985 National Forum for School Science. Washington, D.C.: The American Association for the Advancement of Science.

Charters, W. W., Jr.

1976 Some obvious facts about the teaching career. *Educational Administration Quarterly* 3:182-193.

1970 Some factors affecting teacher survival in school districts. *American Educational Research Journal* 7:1-27.

Chelimsky, Eleanor

1988 *Production and Quality of Education Information.* Testimony before the Subcommittee on Select Education Committee on Education and Labor, House of Representatives, U.S. Congress. April. Washington, D.C.: U.S. General Accounting Office.

Cohen, Michael

1988 Designing state assessment systems. *Phi Delta Kappan.* April:583-588.

College Placement Council

1988 *CPC Salary Survey.* Bethlehem, Pennsylvania: College Placement Council.

Colorado State Department of Education

1986 *Teacher Supply and Demand for K-12 Public School Programs in Colorado: 1985 and Beyond.* Prepared by Roger E. Neppl and Jerry Scezney. Denver, Colorado: Author.

Congressional Budget Office

1983a *Initiatives in Science and Mathematics Education: Issues and Options.* Daniel Koretz, author. Draft paper. April 11. Washington, D.C.: Congressional Budget Office.

1983b *Science and Mathematics Education: Issues in Shaping a Federal Initiative.* March 11. Daniel Koretz, author. Washington, D.C.: Congressional Budget Office.

1985 *Reducing Poverty Among Children.* May. Robertson C. Williams and Gina C. Adams, authors. Washington, D.C.: Congressional Budget Office.

1986 *Trends in Educational Achievement.* April. Daniel Koretz, author. Washington, D.C.: Congressional Budget Office.

1987 *Educational Achievement: Explanations and Implications of Recent Trends.* August. Daniel Koretz, author. Washington, D.C.: Congressional Budget Office.

Congressional Research Service

1988 Education: the challenges. *Major Issue Forum* (includes a section on math-science literacy and productivity). October. Washington, D.C.: Congressional Research Service.

Connecticut Board of Education

1988 *Teacher Supply and Demand in Connecticut: A Second Look.* November. Hartford: Connecticut Board of Education.

Connecticut State Department of Education

1984 *Public School Enrollment Projections Through 2000.* Hartford, Connecticut: Author.

1985a *School Staff Report.* Hartford, Connecticut: Author.

1985b *Teacher Supply and Demand in Connecticut.* Hartford, Connecticut: Author.

1987 A Second Look at Connecticut Teacher Supply and Demand. Part I: A Study of Connecticut's 1985 Teacher Attrition Rates. Unpublished report.

1988 *Fall Hiring Report: Certified Professional Staff Vacancies as of September 1, 1987.* Hartford: State of Connecticut Department of Education.

Connecticut State Department of Higher Education

1983 *Student Migration Into and Out of Connecticut: An Update.* Hartford, Connecticut: Author.

1985 *Degrees Conferred by Connecticut Institutions of Higher Education 1982-83.* Research Report R-1-85. Hartford, Connecticut: Author.

Contra, John A., and David A. Potter.

1980 School and teacher effect: An interrelational model. *Review of Educational Research.* Summer:287.

Council of Chief State School Officers

1984 *Staffing the Nation's Schools: A National Emergency.* Washington, D.C.: Author.

Crane, Jane

1982 *Teacher Demand: A Socio-Demographic Phenomenon.* National Center for Education Statistics. Washington, D.C.: U.S. Department of Education.

Darling-Hammond, Linda

1984 *Beyond the Commission Reports: The Coming Crisis in Teaching.* Santa Monica, California: The RAND Corporation.

1986 *Teacher Supply and Demand: A Structural Perspective.* April. Paper for presentation at 1986 American Educational Research Association Annual Meeting. Santa Monica, California: The RAND Corporation.

1988a The futures of teaching. *Educational Leadership.* November:4-10.

1988b On teaching as a profession: a conversation with Linda Darling-Hammond. (by Anne Meek) *Educational Leadership.* November:11-17.

Darling-Hammond, Linda, Gus Haggstrom, Lisa Hudson, and Jennie Oakes

1986 *A Conceptual Framework for Examining Staffing and Schooling.* Draft report prepared for the Center for Education Statistics. Santa Monica, California: The RAND Corporation.

Darling-Hammond, Linda, and Lisa Hudson

1986 *Indicators of Teacher and Teaching Quality.* Draft report prepared for the National Science Foundation. Santa Monica, California: The RAND Corporation.

1987a *Indicators of Teacher Quality for a Comprehensive Monitoring System of Mathematics and Science Education.* September. Paper prepared for the National Science Foundation project "Monitoring National Progress in Mathematics, Science and Technology Education." Santa Monica, California: The RAND Corporation.

1987b *Precollege Science and Mathematics Teachers: Supply, Demand and Quality.* Santa Monica, California: The RAND Corporation.

Darling-Hammond, Linda, Lisa Hudson, and Sheila Nataraj Kirby

1989 *Redesigning Teacher Education: Opening the Door for New Recruits to Science and Mathematics Teaching.* March. R-3661-FF/CSTP. Santa Monica, California: The RAND Corporation.

David, Jane L.

1988 The use of indicators by school districts: aid or threat to improvement? *Phi Delta Kappan* March:499-502.

Dean, B. V., A. Reisman, and E. Rattner

1971 *Supply and Demand of Teachers and Supply and Demand of Ph.D's 1971-1980.* Revised draft, June. Cleveland, Ohio: Case Western Reserve University.

Delaware State Department of Public Instruction

1985a *Educational Personnel in Delaware Public Schools, 1984-85.* Prepared by Robert F. Boozer. Dover, Delaware: Author.

1985b *Supply and Demand: Educational Personnel in Delaware, 1984-85.* Prepared by Robert F. Boozer. Dover, Delaware: Author.

Dossey, J. A., Ina V. S. Mullis, Mary M. Lindquist, and Donald L. Chambers

1988 *The Mathematics Report Card: Are We Measuring Up? Trends and Achievement Based on the 1986 National Assessment.* Princeton, New Jersey: Educational Testing Service.

Douglas, Stratford, and Ronald E. Bird

1985 *Personal Labor Supply: An Estimation of a Probabilistic Teacher Labor Supply Function.* Occasional Papers in Educational Policy Analysis No. 416. Research Triangle Park, North Carolina: Southeastern Regional Council for Educational Improvement.

Druva, Cynthia A., and Ronald J. Anderson

1983 Science teacher characteristics by teacher behavior and by student outcome: a meta-analysis of research. *Journal of Research in Science Teaching.* 20(5)467-479.

Duggan, Paula
1988 *Labor Market Information: Book I—A State Policymaker's Guide*. Washington, D.C.: Northeast-Midwest Institute: The Center for Regional Policy.

Dworkin, Anthony G.
1987 *Teacher Burnout in the Public Schools: Structural Causes and Consequences for Children*. Albany: State University of New York Press.

Eberts, R., and J. Stone
1984 *Unions and Public Schools*. Lexington, Massachusetts: Lexington Books.

Eccles, J., and Lois W. Hoffman
1985 Sex roles, socialization, and occupational behavior. In H. W. Stevenson and A. E. Siegel, eds., *Research in Child Development and Social Policy*. Vol. I. Chicago: University of Chicago Press.

Education Commission of the States
1984 *A 50-State Survey of Initiatives for Attracting Science and Mathematics Teachers*. Denver, Colorado: Author.
1985 *New Directions for State Teacher Policies*. Working Paper No. TR-85-1. Prepared by Judith L. Bray, Patricia Flakus-Mosqueda, Robert M. Palaich, and JoAnne S. Wilkins. Denver, Colorado: Author.
1987 *Minimum High School Graduation Requirements in 1980, 1985 and 1987: Standard Diplomas*. No. CN28. Denver, Colorado: Author.
1988 State Institutional Initiatives for Minority Teacher Recruitment. By Joy Miller. Working Paper No. TE-88-C1. Denver, Colorado: Author.

Education Week
1985 Changing course: a 50-state survey of reform measures. *Education Week* February 6:11-30.

Educational Testing Service
1984 *Recapturing the Lead in Math and Science. Focus (14)*. Princeton, New Jersey: Educational Testing Service.
1986 *National Report: College-Bound Seniors, 1986. An ATP Summary Report*. Princeton, New Jersey: College Entrance Examination Board.
1987 *1987 Profile of SAT and Achievement Test Takers. A National Report on College-Bound Seniors*. Princeton, New Jersey: College Entrance Examination Board.
1989 *What Americans Study. A Policy Information Report*. Princeton, New Jersey: Educational Testing Service.

Eissenberg, Thomas E., and L. M. Rudner
1988 State testing of teachers: a summary. *Journal of Teacher Education*. July-August:21-22.

Elliott, E. J., and C. D. Cowan
1987 *Redesigning the Collection of National Education Statistics*. Paper. Washington, D.C.: Center for Education Statistics, U.S. Department of Education.

Elsworth, Gerald R., and Frank Coulter
1978 Relationships between professional self-perception and commitment to teaching. *Australian Journal of Education* 22(March):25-37.

Executive Service Corps of New England, Inc.
1989 *High School Math/Science Project—Final Report of Interviews on Using Retirees to Teach Math and Science*. March 8. Boston, Massachusetts: Executive Service Corps of New England, Inc.

Falk, W. W., C. Falkowski, and T. A. Lyson
1981 Some plan to be teachers: further elaboration and specification. *Sociology of Education* 54:64-69.

Feistritzer, C. Emily

1983a *The American Teacher*. Washington, D.C.: Feistritzer Publications.

1983b *The Condition of Teaching: A State by State Analysis*. Princeton, New Jersey: The Carnegie Foundation for the Advancement of Teaching.

1984 *The Making of a Teacher: A Report on Teacher Education and Certification*. Washington, D.C.: National Center for Education Information.

1985 *The Condition of Teaching: A State by State Analysis, 1985*. Princeton, New Jersey: The Carnegie Foundation for the Advancement of Teaching.

1988a *Profile of School Administrators in the U.S.* Washington, D.C.: National Center for Education Information.

1988b *Teacher Supply and Demand Surveys 1988*. Washington, D.C.: National Center for Education Information.

Fielstra, C.

1955 An analysis of factors influencing the decision to become a teacher. *Journal of Educational Research* 48:659-667.

Flakus-Mosqueda, Patricia

1983 *Survey of States' Teacher Policies*. Working Paper No. 2. Denver, Colorado: Education Commission of the States.

Flanders, J.

1986 How Much of the Content in Mathematics Textbooks is New? Unpublished paper, University of Chicago.

Florida State Department of Education

1980 *Trends in Teacher Supply and Demand*. Prepared by Al Fresin. Tallahassee, Florida: Author.

1982 *Areas of Critical Teacher Need in Florida: 1980-81*. Tallahassee, Florida: Author.

1983 *Teacher Supply and Demand in Florida: Second Annual Report*. Tallahassee, Florida: Author.

1984 *Teacher Supply and Demand in Florida: Third Annual Report*. Tallahassee, Florida: Author.

1985 *Teacher Supply and Demand in Florida. Fourth Annual Report*. Tallahassee, Florida: Author.

1986 *Teacher Supply and Demand in Florida: Fifth Annual Report*. October. Tallahassee, Florida: Florida Department of Education.

n.d. *Methodology for the Identification of Teacher Shortage Areas, 1986-87*. Tallahassee, Florida: Author.

Florida State Education Standards Commission

1985 *Teaching as a Career: High School Students' Perceptions of Teachers and Teaching*. Tallahassee, Florida: Author.

1986 *Teachers for Florida's Classrooms: Meeting the Challenge*. Tallahassee, Florida: Author.

Frankel, Martin, and Debra Gerald

1984 Projections of Education Statistics: An Analysis of Projection Errors. National Center for Education Statistics, U.S. Department of Education, Washington, D.C.

Freeman, Richard B.

1971 *The Market for College-Trained Manpower*. Cambridge, Massachusetts: Harvard University Press.

1975 Legal cobwebs: a recursive model of the market for new lawyers. *Review of Economics and Statistics* 57(2):171-179.

Futrell, Mary H.

1987 *Statement Before the 125th Annual NEA Convention Opening News Conference.* July 1. Washington, D.C.: National Education Association.

Galambos, Eva C.

1980 *The Changing Labor Market for Teachers in the South.* Atlanta: Southern Regional Education Board.

1985 *Teacher Preparation: The Anatomy of a College Degree.* Atlanta: Southern Regional Education Board.

1986 *Teachers in the United States.* Report for the U.S. Department of Education. March 28 draft. Atlanta, Georgia: Author.

Garden, Robert A.

1987 *Second IEA Mathematics Study: Sampling Report.* March. Center for Education Statistics. Washington, D.C.: U.S. Department of Education.

Gilford, Dorothy M.

1987 Data Resources to Describe U.S. Precollege Science and Mathematics Curricula. In A. B. Champagne and L. E. Hornig, eds, *The Science Curriculum: The Report of the 1986 National Forum for School Science.* Washington, D.C.: American Association for the Advancement of Science.

Gilford, Dorothy M., and Ellen Tenenbaum

1985 *Report of the Planning Conference for a Study of Statistics on Supply of and Demand for Precollege Science and Mathematics Teachers.* Committee on National Statistics. Washington, D.C.: National Research Council.

Ginsberg, R., and R. K. Wimpelberg

1987 Educational change by commission: attempting "trickle down" reform. *Educational Evaluation and Policy Analysis* 9(4)(winter):344-360.

Ginsberg, Rick, Elchanan Cohn, C. Glyn Williams, S. Travis Pritchett, and Tom Smith

1989 *Teaching in South Carolina: A Retirement Initiative.* South Carolina Educational Policy Center, College of Education. Columbia: University of South Carolina.

Glass, G. V., L. S. Cahen, M. L. Smith, and N. N. Filby

1984 *School Class Size: Research and Policy.* Beverly Hills, California: Sage Publications.

Goertz, Margaret E.

1985 *State Educational Standards: A 50-State Survey.* Princeton, New Jersey: Educational Testing Service.

1989 *Course-Taking Patterns in the 1980s.* Princeton, New Jersey: Educational Testing Service.

Goertz, Margaret E., Ruth B. Ekstrom, and Richard J. Coley

1984 *The Impact of State Policy on Entrance into the Teaching Profession.* Report to National Institute of Education. Princeton, New Jersey: Educational Testing Service.

Goodlad, John I.

1984 *A Place Called School.* New York: McGraw-Hill.

Garden, Robert A.

1987 *Second IEA Mathematics Study: Sampling Report.* March. Center for Education Statistics. Washington, D.C.: U.S. Department of Education.

Graham, Patricia A.

1987 Black teachers: a drastically scarce resource. *Phi Delta Kappan* April:598-605.

Grant Foundation Commission on Work, Family and Citizenship

1988 *The Forgotten Half: Pathways to Success for America's Youth and Young Families*. November. Washington, D.C.: The William T. Grant Commission on Work, Family and Citizenship.

Graybeal, William S.

1979 *Teacher Supply and Demand in Public Schools in 1978*. Washington, D.C.: National Education Association.

1980 *Teacher Supply and Demand in Public Schools in 1979*. Washington, D.C.: National Education Association.

1981 *Teacher Supply and Demand in Public Schools in 1980-81*. Washington, D.C.: National Education Association.

1983 *Teacher Supply and Demand in Public Schools in 1981-82*. Washington, D.C.: National Education Association.

Greenberg, D., and J. McCall

1973 *Analysis of the Educational Personnel System: I. Teacher Mobility in San Diego*. Santa Monica, California: The RAND Corporation.

Grissmer, David W.

1985 Projections of Teacher Demand for the District of Columbia Schools. Draft report. Santa Monica, California: The RAND Corporation.

Grissmer, David W., and Linda Darling-Hammond

1984 *A Prototype Personnel Planning System for the D.C. Public Schools*. Santa Monica, California: The RAND Corporation.

Grissmer, David W. and Sheila N. Kirby

1987 *Teacher Attrition: The Uphill Climb to Staff the Nation's Schools*. Santa Monica, California: The RAND Corporation.

Guthrie, James W.

1985 The educational policy consequences of economic instability: the emerging political economy of American education. *Educational Evaluation and Policy Analysis* 7(4)(Winter):319-332.

Guthrie, James W., and Ami Zusman

1982a *Mathematics and Science Teacher Shortages: What Can California Do?* Berkeley: Institute of Governmental Studies, University of California.

1982b Teacher supply and demand in mathematics and science. *Phi Delta Kappan* 64(September):28-33.

Haberman, Martin

1988 Proposals for recruiting minority teachers: promising practices and attractive detours. *Journal of Teacher Education*. July/August:38-44.

Haggstrom, Gus W., Linda Darling-Hammond, and David W. Grissmer

1986 *Assessing Teacher Supply and Demand*. Draft report prepared for the Center for Education Statistics. Santa Monica, California: The RAND Corporation.

1988 *Assessing Teacher Supply and Demand*. May. Santa Monica, California: The RAND Corporation.

Haney, Walt, George Madaus, and Amelia Kreitzer

1987 Charms talismanic: testing teachers for the improvement of American education. In Ernst Rothkopf, ed., *Review of Research in Education*. Washington, D.C.: American Educational Research Association.

Hanson, David

1988 Study gives low grades to U.S. precollege science education. *Chemical and Engineering News* October 10.

Hanushek, Eric A.

1986 The economics of schooling: production and efficiency in public schools. *Journal of Economic Literature* 24:1141-1177.

1989 The impact of differential expenditures on school performance. *Educational Researcher* May:45-51.

Harris, Seymour

1949 *The Market for College Graduates.* Cambridge, Massachusetts: Harvard University Press.

Haubrich, V. F.

1960 The motives of prospective teachers. *Journal of Teacher Education* 11:381-386.

Hausman, H. J., and A. H. Livermore

1978 *Supply and Demand for High School Science Teachers in 1985.* Columbus, Ohio: ERIC Clearinghouse for Science, Mathematics and Environmental Education.

Heikkinen, M. W.

1988 The academic preparation of Idaho science teachers. *Science Education* 72(1):63-71.

Hewson, P. W., and M. G. Hewson

1987 Science teachers' conceptions of teaching: implications for teacher education. *International Journal of Science Education* 9(4):425-440.

Heyns, Barbara

1988 Educational defectors: a first look at teacher attrition in the NLS-72. *Educational Researcher* April:24-32.

Hill, C. R., and F. P. Stafford

1985 Parental care of children: Time diary estimates of quantity, predictability and variety. In F. T. Juster and F. P. Stafford, eds. *Time, Goods, and Well-Being.* Ann Arbor, Michigan: Institute for Social Research.

Hill, M. S., S. Augustyniak, and M. Ponza

1987 *Effects of Parental Divorce on Children's Attainments: An Empirical Comparison of Five Hypotheses.* September. Ann Arbor, Michigan: Institute for Social Research.

Hill, Susan T.

1983 *Participation of Black Students in Higher Education: A Statistical Profile from 1970-71 to 1980-81.* November. Washington, D.C.: National Center for Education Statistics.

Holden, Constance

1989 Wanted: 675,000 future scientists and engineers. *Science* 244(June 30):1536-1537.

Holland, John L.

1976 Vocational preferences. Pp. 521-570 in Marvin D. Dunnette, editor, *Handbook of Industrial and Organizational Psychology.* Chicago, Illinois: Rand McNally College Publishing Company.

Holmes Group, Inc.

1986 *Tomorrow's Teachers: A Report of the Holmes Group.* East Lansing, Michigan: Author.

Holmstrom, E. I.

1985 *Recent Changes in Teacher Education Programs.* Washington, D.C.: American Council on Education.

Hopfengardner, J. D., T. Lasley, and E. Joseph
1983 Recruiting preservice teacher education students. *Journal of Teacher Education* 34:10-13.

Howe, Trevor G., and Jack A. Gerlovich
1982 *National Study of the Estimated Supply and Demand of Secondary Science and Mathematics Teachers*. Ames, Iowa: Iowa State University.

Hudson, L., and L. Darling-Hammond
1987 *The Schooling and Staffing Surveys: Analysis and Recommendations*. Report No. WD-3414-ED. April. Santa Monica, California: The RAND Corporation.

Hudson, L., and S. N. Kirby
1988 A Profile of Nontraditional Recruits to Mathematics and Science Teaching. Paper presented at the annual meeting of the American Educational Research Association, New Orleans, Louisiana. April.

Hudson, Lisa, Sheila Naturuj Kirby, Neil B. Carey, Brian S. Mittman, and Barnett Berry
1988 *Recruiting Mathematics and Science Teachers Through Nontraditional Programs: Case Studies*. A RAND note, August. Santa Monica, California: The RAND Corporation.

Hueftle, S. J., S. J. Rakow, and W. U. Welch
1983 *Images of Science: Summary Results from the 1981-82 National Assessment in Science*. Minneapolis, Minnesota: Minnesota Research and Evaluation Center.

Illinois State Board of Education
1983 *The Supply and Demand for Illinois Mathematics and Science Teachers*. Springfield, Illinois: Author.
1984a *Average Starting Salaries of Selected Professionals, 1984*. Springfield, Illinois: Author.
1984b *Illinois Teacher Supply and Demand. 1983-84 Summary and Data Tables*. Springfield, Illinois: Author.
1984c *Vacant Educational Positions in Illinois School Districts*. Springfield, Illinois: Author.
1985a *Illinois Public School Enrollment Analysis and Projections: 1985*. Springfield, Illinois: Author.
1985b *Illinois Teacher Supply and Demand, 1984-1985*. Springfield, Illinois: Author.
1989 *Illinois Teacher Supply and Demand, 1986-87*. Springfield, Illinois: Author.

Institute of Electrical and Electronics Engineers
1984 National crisis in pre-college science and math education. *Impact* 8(3)(May).

International Association for the Evaluation of Educational Achievement (IEA)
1988 *Science Achievement in Seventeen Countries: A Preliminary Report*. New York: Pergamon Press.

Jacobson, Willard J., and Rodney L. Doran
1988 *Science Achievement in the United States and Sixteen Countries: A Report to the Public*. New York: Second International Science Study, Teachers College, Columbia University.

Jamar, D., and S. Ervay
1983 The effect of teacher education on the career goals of women. *Phi Delta Kappan* 65:593.

Johnson Foundation
1988 *Conversations from Wingspread: What Teachers Ought to Know; Preparing Teachers; Minority Teacher Education; Minority Teachers; Understanding*

Science; Today's Schools and Tomorrow's Workforce. Racine, Wisconsin: The Johnson Foundation.

Johnston, K. L., and B. G. Aldridge
1984 The crisis in science education: What is it? How can we respond? *Journal of College Science Teaching*. September-October:20-28.

Jump, Bernard, Jr.
1986 Teacher Mobility and Pension Portability. Paper prepared for the Carnegie Forum on Education.

Kemple, James J.
1989 The Career Paths of Black Teachers: Evidence from North Carolina. Paper presented at the annual meeting of the American Educational Research Association, San Francisco, California. March. Graduate School of Education, Harvard University.

Kentucky State Council on Higher Education
1980 *A Study of Teacher Education in Kentucky*. Prepared by Harry M. Snyder. Frankfort, Kentucky: Author.

Kirst, Michael W.
1987 *Proposal for Analysis, Reporting, and Dissemination Activities Concerning the Conditions of Children in America*. April. Palo Alto, California: Stanford University PACE Office.

Kotlikoff, Laurence J., and Danile E. Smith
1983 *Pensions in the American Economy*. National Bureau of Economic Research Monograph. Chicago: University of Chicago Press.

Krein, Shirley F., and Andrea H. Beller
1988 Educational attainment of children from single-parent families: differences by exposure, gender, and race. *Demography* 25(2)(May):122.

Lanier, Judith, and Joseph Featherstone
1988 A new commitment to teacher education. *Educational Leadership* 46(3)(November).

Lashier, William S., and Wan Yung Ryoo
1984 A longitudinal study of the supply and demand for physics and chemistry teachers in Kansas. *Journal of Research in Science Teaching* 21(1):17-26.

Lee, Shin-ying, Veronica Ichikawa, and Harold W. Stevenson
1987 Beliefs and achievements in mathematics and reading: a cross-national study of Chinese, Japanese, and American children and their mothers. *Advances in Motivation and Achievement: Enhancing Motivation* 5:149-179. Greenwich, Connecticut: JAI Press, Inc.

Levin, Henry L.
1979 Educational production theory and teacher inputs. Chapter 5 in R. Dreeben and J. A. Thomas, eds., *The Analysis of Educational Productivity: Issues in Microanalysis*. Cambridge, Massachusetts: Ballinger.

Levin, Henry M.
1985 Solving the shortage of mathematics and science teachers. *Educational Evaluation and Policy Analysis* 7(4):371-382.

Lindenfield, F.
1963 *Teacher Turnover in Public Elementary and Secondary Schools, 1959-1960*. Circular 675. Washington, D.C.: Office of Education, U.S. Department of Health, Education and Welfare.

Litt, M.S., and D.C. Turk
1983 Stress, Dissatisfaction, and Intention to Leave Teaching in Experienced

Public High School Teachers. Paper presented to the American Educational Research Association Annual Meeting, Montreal.

Locke, Edwin A.
1976 The nature and causes of job satisfaction. In Marvin D. Dunnette, ed., *Handbook of Industrial and Organizational Psychology*. Chicago: Rand McNally.

Louisiana State Department of Education
1982 *Declining Teacher Supply*. Baton Rouge, Louisiana: Author.

Manski, Charles F.
1985 *Academic Ability, Earnings, and the Decision to Become a Teacher: Evidence from the National Longitudinal Study of the High School Class of 1972.* Working paper No. 1539. New York: National Bureau of Economic Research.
1987 Academic ability, earnings, and the decision to become a teacher: evidence from the National Longitudinal Study of the High School Class of 1972. In David A. Wise, ed., *Public Sector Payrolls*. Chicago: University of Chicago Press.

Mark, J. H., and B. D. Anderson
1985 Teacher survival rates in St. Louis, 1969-1982. *American Educational Research Journal* 22(3):413-421.

Marrett, Cora B.
1987 Black and Native American students in precollege mathematics and science. In Linda S. Dix, ed. *Minorities: Their Underrepresentation and Career Differentials in Science and Engineering.* Workshop on Minorities in Science and Engineering, Nov. 21, 1986. Washington, D.C.: National Academy Press.

Maryland State Department of Education
n.d. *Supply and Demand for Mathematics and Science Teachers, 1985-1990.* Annapolis, Maryland: Office of Research and Evaluation.
1988 *Teacher Supply and Demand in Maryland, 1987-1988*. September. Annapolis, Maryland: Author.

Maryland State Department of Education and Maryland State Board for Higher Education
1985 *Teacher Supply and Demand in Maryland 1985-1987*. Annapolis, Maryland: Authors.
1987 *Teacher Supply and Demand in Maryland 1987-1990*. September. Annapolis, Maryland: Authors.

Massachusetts Institute for Social and Economic Research (MISER)
1987 *Report on the Status of Teacher Supply and Demand in Massachusetts*. June 1. Amherst, Massachusetts: University of Massachusetts.

Mathematica Policy Research, Inc.
1986 *A Review of Teacher Supply and Demand Projections by the U.S. Department of Education, Illinois, and New York*. August. Cavin, E., author. Princeton, New Jersey: Author.

Mathematical Association of America
1983 *Recommendations on the Mathematical Preparation of Teachers*. CUPM Panel on Teacher Training. Washington, D.C.: Author.

Maxwell, James P.
1986 The Academic Achievement of Freshman and Junior Education Candidates. Paper presented at the annual meeting of the American Educational Research Association. April.

McKibbin, Michael D.
1988 Alternative teacher certification programs. *Educational Leadership* November: 32-35.

McKnight, C. C., F. J. Crosswhite, J. A. Dossey, E. Kifer, J. D. Swafford, K. K. Travers, and T. J. Cooney
1987 *The Underachieving Curriculum: Assessing U.S. School Mathematics from an International Perspective*. Champaign, Illinois: Stipes Publishing Company.

Metropolitan Life Insurance Company
1988 *The American Teacher—Metropolitan Life Survey*. Conducted by Louis Harris and Associates, Inc. New York: Metropolitan Life Insurance Company.

Metz, A. S., and H. L. Fleischman
1974 *Teacher Turnover in Public Schools Fall 1968 to Fall 1969*. U.S. Department of Health, Education, and Welfare, DHEW Publication No. (OE) 74-11115. Washington, D.C.: U.S. Government Printing Office.

Metz, A. Stafford, and John P. Sietsma
1982 *Teachers Employed in Public Schools 1979-80*. NCES Special Report, May. Washington, D.C.: U.S. Department of Education.

Mississippi State Department of Education
1979 *Teacher Supply and Demand in Mississippi*. Prepared by James H. Jones. Jackson, Mississippi: Author.

Mitchell, Charles, and Joyce Maar
1987 Elementary school certification practice: Is it time for a change? *Education* 107(3):267-273.

Murnane, Richard J.
1980 Interpreting the Evidence on School Effectiveness. Working Paper No. 830. March 24. New Haven, Connecticut: Institution for Social and Policy Studies, Yale University.
1981a *Interpreting the evidence on school effectiveness*. Teachers College Record. 83(1)(Fall)1981:19-30.
1981b Teacher mobility revisited. *Journal of Human Resources* 16(1):3-16.
1983 Selection and survival in the teacher labor market. *The Review of Economics and Statistics*:513-518.

Murnane, Richard J., and Randall J. Olsen
1987 How Long Teachers Stay in Teaching Depends on Salaries and Opportunity Costs. December. Paper presented at the December 1987 Econometrics Society Meetings, Chicago, Illinois.
1989a The effects of salaries and opportunity costs on duration in teaching: evidence from Michigan. *Review of Economics and Statistics* Forthcoming.
1989b Will there be enough teachers? *American Economic Review Papers and Proceedings*. May.
1990 The effects of salaries and opportunity costs on length of stay in teaching: evidence from North Carolina. *Journal of Human Resources* Forthcoming.

Murnane, Richard J., and Barbara R. Phillips
1981 Learning by doing, vintage and selection: three pieces of the puzzle relating teaching experience and teaching performance. *Economics of Education Review* 1(4):453-465.

Murnane, Richard J., and Senta A. Raizen, eds.
1988 *Improving Indicators of the Quality of Science and Mathematics Education in*

Grades K-12. Committee on Indicators of Precollege Science and Mathematics Education, Commission on Behavioral and Social Sciences and Education. Washington, D.C.: National Academy Press.

Murnane, Richard J., and Michael Schwinden
1989 Race, gender, and opportunity: supply and demand for new teachers in 1975-85. *Educational Evaluation and Policy Analysis* Summer 11(2):93-108.

Murnane, Richard J., Judith D. Singer, and John B. Willett
1988 The career paths of teachers: implications for teacher supply and methodological lessons for research. *Educational Researcher* August-September:22-30.
1989 The influences of salaries and "opportunity costs" on teachers' career choices: evidence from North Carolina. *Harvard Educational Review*. August:59(3):325-346.

National Academy of Engineering
1989 *Technology and Competitiveness: A Statement of the Council of the National Academy of Engineering*. Supplement to *The Bridge*. 19(1)(spring). Washington, D.C.: National Academy of Engineering.

National Academy of Sciences
1987 *Nurturing Science and Engineering Talent: A Discussion Paper*. July. Government-University-Industry Research Roundtable. Washington, D.C.: National Academy of Sciences.

National Academy of Sciences and National Academy of Engineering
1982 *Science and Mathematics in the Schools: Report of a Convocation*. Washington, D.C.: National Academy Press.

National Assessment of Educational Progress (NAEP)
1985 *The Reading Report Card: Trends in Reading over Four National Assessments, 1971-1984*. Princeton, New Jersey: Educational Testing Service.
1986 *The Writing Report Card*. Princeton, New Jersey: Educational Testing Service.
1987 *Implementing the New Design: The NAEP 1983-84 Technical Report*. Report No. 15-TR-20. Princeton, New Jersey: Educational Testing Service.
1988a *The Mathematics Report Card: Are We Measuring Up? Trends and Achievement Based on the 1986 National Assessment*. Report No. 17-M-01. Princeton, New Jersey: Educational Testing Service.
1988b *Science Learning Matters: An Overview of the Science Report Card*. September. Princeton, New Jersey: Educational Testing Service.
1988c *The Science Report Card: Trends and Achievement Based on the 1986 National Assessment*. Report No. 17-S-01. Princeton, New Jersey: Educational Testing Service.

National Association of Secondary School Principals
1985 *An Agenda for Excellence at the Middle Level*. September. Reston, Virginia: National Association of Secondary School Principals.

National Association of State Directors of Teacher Education and Certification
1988 *Manual on Certification*. Sacramento, California: National Association of State Directors of Teacher Education and Certification.

National Center for Education Statistics (NCES)
1982a *Projections of Education Statistics to 1990-91*. Volume 2: *Methodological Report*. By M. Frankel and D. Gerald. Washington, D.C.: U.S. Government Printing Office.
1982b *Teachers Employed in Public Schools 1979-80*. A report on the 1979-80 Survey of Teacher Demand and Shortage. Prepared by S. A. Metz and J. P. Sietsma. Washington, D.C.: U.S. Department of Education.

1983 *New Teachers in the Job Market: 1981 Update*. Prepared by Jane L. Crane. Washington, D.C.: U.S. Government Printing Office.

1984a *The Condition of Education*. 1984 edition. Washington, D.C.: U.S. Department of Education.

1984b *Projections of Education Statistics: An Analysis of Projection Errors*. Prepared by M. Frankel and D. Gerald. September. Washington, D.C.: U.S. Department of Education.

1984c Science and mathematics education in American high schools: results from the High School and Beyond Study. *NCES Bulletin*. May. Washington, D.C.: U.S. Department of Education.

1985a *The Condition of Education*. 1985 edition. Washington, D.C.: U.S. Department of Education.

1985b *Projections of Education Statistics to 1992-93: Methodological Report with Detailed Projection Tables*. Prepared by Debra E. Gerald. Washington, D.C.: U.S. Department of Education.

1988a *Background and Experience. Characteristics of Public and Private School Teachers: 1984-85 and 1985-86, Respectively*. October. Survey Report CS 88-1103. Washington, D.C.: National Center for Education Statistics.

1988b *Digest of Education Statistics 1988*. Washington, D.C.: U.S. Department of Education.

1988c *E.D.TABS: Postsecondary Fall Enrollment, 1986*. Schools and Staffing Survey (SASS). July. Washington, D.C.: U.S. Department of Education, Office of Education Research and Improvement.

1988d *Report of the Conference on Elementary and Secondary Education Statistics Program*. January 25-26. Report by S. A. Raizen. Washington, D.C.: U.S. Department of Education.

1988e *Trends in Minority Enrollment in Higher Education, Fall 1976-Fall 1986*. Survey Report CS 88-201. April. Washington, D.C.: U.S. Department of Education.

1988f *Education Indicators—1988*. Report CS88-624. Washington, D.C.: U.S. Government Printing Office..

1988g *Projections of Education Statistics to 1997-98*. CS 88-607. Washington, D.C.: U.S. Department of Education.

1989a *E.D.TABS: Schools and Staffing Survey*. Washington, D.C.: U.S. Department of Education, Office of Education Research and Improvement.

1989b *Minority Student Issues: Racial/Ethnic Data Collected by the National Center for Education Statistics Since 1969*. March. CS 89-267. Washington, D.C.: U.S. Department of Education.

1989c *State Projections to 1993 for Public Elementary and Secondary Enrollment, Graduates, and Teachers*. Report No. CS 89-638. Washington, D.C.: U.S. Department of Education.

National Commission on Excellence in Education

1983 *A Nation at Risk: The Imperative for Educational Reform*. Washington, D.C.: U.S. Department of Education.

National Education Association (NEA)

1957 *Status of American School Teachers*. Washington, D.C.: Author.

1972 *Status of the American Public School Teacher, 1970-1971*. Washington, D.C.: Author.

1982 *Status of the American Public School Teacher, 1980-1981*. Washington, D.C.: Author.

1987a NEA Poll Says Teachers Better Educated, Have More Experience and Work Longer Hours. Press release. July 1. National Education Association, Washington, D.C.

1987b Nearly Half of Teachers Report Overcrowded Classrooms. Press release. July 1. National Education Association, Washington, D.C.

1987c Number of Misassigned Teachers Rising, New NEA Survey Discloses. Press release. July 1. National Education Association, Washington, D.C.

1987d Percentage of Minority Teachers Declines as Minority Enrollment Increases. Press release. July 1. National Education Association, Washington, D.C.

1987e *Status of the American Public School Teacher: 1985-86.* Washington, D.C.: National Education Association.

1987f *Teacher Supply and Demand Guidebook.* Washington, D.C.: National Education Association.

1987g Teachers Forced to Moonlight to Supplement Family Income. Press release. July 1. National Education Association, Washington, D.C.

1987h Teachers Spend Average of 11 Hours Weekly on School-Related Activities Without Pay. Press release. July 1. National Education Association, Washington, D.C.

1988 NEA Statement on Teacher Shortage. Press release. September 16. National Education Association, Washington, D.C.

National Governors' Association

1986 *Time for Results: The Governors' 1991 Report on Education.* August. Washington, D.C.: Author.

1987 *Results in Education: 1987.* Washington, D.C.: Author.

1988 *Pension Portability for Educators: A Plan for the Future.* By Jean G. McDonald. Washington, D.C.: Author.

National Research Council

1977 *Fundamental Research and the Process of Education.* Committee on Fundamental Research Relevant to Education, Assembly of Behavioral and Social Sciences. Washington, D.C.: National Academy of Sciences.

1979 *The State of School Science.* June. Panel on School Science, Commission in Human Resources. Washington, D.C.: National Academy of Sciences.

1985 *Indicators of Precollege Education in Science and Mathematics: A Preliminary Review.* Committee on Indicators of Precollege Science and Mathematics Education. Washington, D.C.: National Academy Press.

1986 *Creating a Center for Education Statistics: A Time for Action.* Panel to Evaluate the National Center for Education Statistics, Committee on National Statistics, Commission on Behavioral and Social Sciences and Education. Washington, D.C.: National Academy Press.

1987a *Annual Report of the Mathematical Sciences Education Board.* June. Papers on curriculum reform included in folder. Washington, D.C.: National Academy of Sciences.

1987b *Conference on the Teaching Profession.* Papers include a discussion of NCTM's *Curriculum and Evaluation Standards* by Thomas Cooney and the 1987 *Annual Report of the Mathematical Sciences Education Board (1987)* is also included. Washington, D.C.: National Research Council.

1987c *Toward Understanding Teacher Supply and Demand: Priorities for Research and Development.* Panel on Statistics on Supply and Demand for Precollege Science and Mathematics Teachers, Committee on National Statistics and Committee on Indicators of Precollege Science and Mathematics Education,

Commission on Behavioral and Social Sciences and Education. Washington, D.C.: National Academy Press.

1987d *Interdisciplinary Research in Mathematics, Science, and Technology Education.* Committee on Research in Mathematics, Science, and Technology Education, Commission on Behavioral and Social Sciences and Education. Washington, D.C.: National Academy Press.

1987e *Minorities: Their Underrepresentation and Career Differentials in Science and Engineering—Proceedings of a Workshop.* Office of Scientific and Engineering Personnel. Washington, D.C.: National Academy Press.

1987f *The Teacher of Mathematics: Issues for Today and Tomorrow.* Proceedings of a Conference. October 1987. Cosponsored by the Mathematical Sciences Education Board and the Center for Academic Interinstitutional Programs. Washington, D.C.: National Academy Press.

1989a *Everybody Counts: A Report to the Nation on the Future of Mathematics Education.* Committee on the Mathematical Sciences in the Year 2000, Board on Mathematical Sciences, Mathematical Sciences Education Board. Washington, D.C.: National Academy Press.

1989b *U.S. School Mathematics from an International Perspective: A Guide for Speakers.* Mathematical Sciences Education Board. Washington, D.C.: National Academy Press.

National Science Board

1982 *Today's Problems, Tomorrow's Crises.* A Report of the National Science Board Commission on Precollege Education in Mathematics, Science and Technology. Washington, D.C.: National Science Foundation.

1983 *Educating Americans for the 21st Century.* Commission on Precollege Education in Mathematics, Science and Technology. Washington, D.C.: National Science Foundation.

1987 *Science and Engineering Indicators—1987.* Report No. NSB 87-1. Washington, D.C.: U.S. Government Printing Office.

National Science Foundation

1983 *Educating Americans for the 21st Century.* National Science Board Commission on Precollege Education in Mathematics, Science and Technology. Washington, D.C.: Author.

1985 *Science Indicators—The 1985 Report.* National Science Board Committee on Science Indicators. Washington, D.C.: Author.

1987 *Directory of NSF-Supported Teacher Enhancement Projects.* Pamphlet NSF 87-21. Washington, D.C.: National Science Foundation.

National Science Resources Center

1988 *Science for Children: Resources for Teachers.* Washington, D.C.: National Academy Press.

National Science Teachers Association

1981 *What Research Says to the Science Teacher.* NSTA No. 471-14776. Washington, D.C.: National Science Teachers Association.

1982 Course offerings in science and mathematics. Table in *A Comparison of Public and Private Schools, 1982-1983.* Washington, D.C.: National Science Teachers Association.

1983a Recommended standards for the preparation and certification of teachers of science at the elementary and middle/junior high school levels: science preparation for preservice elementary teachers. *Journal of College Science Teaching* November:122-127.

1983b *Science Preparation for Preservice Elementary Teachers. Recommended Standards for the Preparation and Certification of Teachers of Science at the Elementary and Middle/Junior High School Levels.* Washington, D.C.: National Science Teachers Association.

1984 Recommended standards for the preparation and certification of secondary school teachers of science: high school science teacher preparation. *The Science Teacher* December:57-62.

1988a *Essential Changes in Secondary School Science: Scope, Sequence and Coordination.* By Bill G. Aldridge. Washington, D.C.: National Science Teachers Association.

1988b *NSTA Certification Requirements for Elementary and High School Science Teachers.* Washington, D.C.: National Science Teachers Association.

New Jersey State Department of Higher Education and Department of Education

1983 *Report of the Advisory Council on Math/Science Teacher Supply and Demand.* Trenton, New Jersey: Author.

New York City Board of Education

1987 *Partners in Teaching: NYC Mentor Teacher Internship Program: The First Year, 1986-1987.* New York: New York City Board of Education, and United Federation of Teachers.

1988 *Mentor Teacher Internship Program, 1986-87.* Prepared by the Office of Educational Assessment. January. New York: Author.

1988 *Science Relicensing Program.* New York: Author.

New York State Education Department

1981 *Report of a Validation Study of Information Contained in New York State's Basic Educational Data System.* Albany, New York: Author.

1982 *The Supply and Recruitment of Persons Certified for Public School Service.* Albany, New York: The State Education Department.

1983a *Age and Retention of Teachers 1967-68 through 1981-82.* Albany, New York: The State Education Department.

1983b *Teachers in New York State—1968 to 1982.* Albany, New York: The State Education Department.

1984 *Comparison of Projected and Actual FTE Classroom Teachers.* Albany, New York: Author.

1985a *Projected Age Distribution and Annual Need for Classroom Teachers.* Files of the Information Center on Education. Albany, New York: The State Education Department.

1985b *Projections of Public and Nonpublic School Enrollment and High School Graduates to 1994-95, New York State.* Albany, New York: The State Education Department.

1985c *Teachers in New York State, 1968 to 1984.* Albany, New York: Author.

1986a *Historical and Projected Pupil/Teacher Ratios for Public Schools.* Albany, New York: Author.

1986b *Teacher Compensation in New York State, 1974-75 and 1984-85 School Years.* Albany, New York: Author.

1987 First Year Teachers as a Percent of New Teacher Hires by Type of District and Subject Area. Unpublished report.

1988 Files of the Information Center on Education, compiled by J. Stiglmeier.

1989 Files of the Information Center on Education, compiled by J. Stiglmeier.

n.d.a Methodology for Projections of Public School Classroom Teacher Needs for Major Subject Areas. Appendix A.

n.d.b *Projections of Public School Classroom Teachers, New York State, 1987-88 to 1991-92*. Albany: State University of New York.

North Carolina State Department of Public Instruction

1980 *Teacher Supply and Demand in North Carolina*. Prepared by H. M. Eldridge. Raleigh, North Carolina: Author.

1982 *North Carolina Science Teacher Profile*. Prepared by William E. Spooner. Raleigh, North Carolina: Author.

Northeast-Midwest Institute

1988 Mapping the Labor Market Information Field. A chart. Washington, D.C.: Northeast-Midwest Institute.

Office of Technology Assessment

1988 *Elementary and Secondary Education for Science and Engineering*. December. Number OTA-TM-SET-41. Washington, D.C.: U.S. Government Printing Office.

1989 *Higher Education for Science and Engineering*. March. OTA-BP-SET-52. Washington, D.C.: U.S. Government Printing Office.

Oklahoma State Department of Education

1982 *Teacher Supply and Demand in Oklahoma Public Schools, 1981-82*. Prepared by Sandra Mayfield. Oklahoma City, Oklahoma: Author.

Oregon State System of Higher Education

1985 *Profile of 1983-84 Graduates of Oregon Schools and Colleges of Education One Year After Graduation*. Eugene, Oregon: Author.

Oregon Teacher Standards and Practices Commission

n.d. *Averting a Teacher Crisis in Oregon*. Prepared by David V. Myton. Salem, Oregon: Author.

Pelavin, Sol, Elizabeth R. Reisner, and Gerry Hendrickson

n.d. Analysis of the National Availability of Mathematics and Science Teachers. Submitted to the Office of Planning, Budget, and Evaluation, U.S. Department of Education. Washington, D.C.: Pelavin Associates, Inc.; Policy Studies Associates, Inc.

Pennsylvania State Department of Education

1979 *Methodology for Projecting Enrollments*. By John Senier. Harrisburg, Pennsylvania: Author.

1983a *Investigating Mathematics and Science Teacher Supply and Demand in Pennsylvania—A Synthesis of PDE Data*. By Grace E. Laverty. Harrisburg, Pennsylvania: Author.

1983b *Science and Mathematics Teacher Supply and Demand and Educational Needs Analysis: A Pennsylvania Report*. By James P. Dorwart. Harrisburg, Pennsylvania: Author.

1985 *Projections—Selected Education Statistics for Pennsylvania to 1989-90*. Harrisburg, Pennsylvania: Author.

Peterson, Paul

1985 Economic and Policy Trends Affecting Teacher Effectiveness in Mathematics and Science. Paper prepared for the American Association for the Advancement of Science. Washington, D.C.: The Brookings Institution.

Popkin, Joel, and B. K. Atrostic

1986 *Evaluation of Models of the Supply and Demand for Teachers*. Paper prepared for the Panel on Statistics on Supply and Demand for Precollege Science and Mathematics Teachers, Committee on National Statistics. Washington, D.C.: Joel Popkin and Company.

Prowda, Peter M., and Barbara Q. Beaudin
1988 New Flow Into Connecticut's Teacher Reserve Pool: A Study of the State's Non-Teaching New 1985-86 Certificants. April. Paper presented at the annual meeting of the American Educational Research Association.

Prowda, Peter M., and David W. Grissmer
1986 A State's Perspective on Teacher Supply and Demand: There is No General Shortage. Paper presented at the annual meeting of the American Educational Research Association.

Raizen, Senta A.
1986 Estimates of Teacher Demand and Supply and Related Policy Issues. Paper presented to the American Educational Research Association Annual Meeting, San Francisco.

Rattner, Edward, Burton V. Dean, and Arnold Reisman
1971 Supply and Demand of Teachers and Supply and Demand of Ph.D.'s. Revised draft. Case Western Reserve University.

The Regional Laboratory for Educational Improvement of the Northeast and Islands
1987a *Proceedings: Teacher Supply and Demand Issues for the Northeast. Symposium with the Chief State School Officers*. September 11. Andover, Massachusetts: Author.
1987b *Teacher Quality: An Issue Brief.* By Anne E. Newton. March. Andover, Massachusetts: Author.

Richards, R.
1960 Prospective students' attitudes towards teaching. *Journal of Teacher Education* 11:375-380.

Richardson-Koehler, Virginia, sr. ed.
1987 *Educators' Handbook: A Research Perspective*. White Plains, New York: Longman, Inc.

Roberson, S. D., T. Z. Keith, and E. R. Page
1983 Now who aspires to teach? *Educational Researcher* 12:13-21.

Robinson, Virginia
1985 Out-of-field teaching: barrier to professionalism. *American Educator* Winter:18-23.

Rock, Donald, Ruth Ekstrom, Margaret Goertz, Thomas Hitton, and Judith Pollack
1985 *Factors Associated with Decline of Test Scores of High School Seniors, 1972-1980.* Report No. CS 85-217, Washington, D.C.: Center for Statistics.

Roth, Robert A.
1981 Comparison of methods and results of major teacher supply and demand studies. *Journal of Teacher Education* 32(Nov.-Dec.): 43-46.

Rumberger, Russell
1985 The shortage of mathematics and science teachers: a review of the evidence. *Educational Evaluation and Policy Analysis* 7(4):355-369.

Sanchez, Rene
1989 D.C. poll cites stress on teachers. *The Washington Post*, June 16, p. C1.

Schlechty, Phillip C., and Victor S. Vance
1981 Do academically able teachers leave education? the North Carolina case. *Phi Delta Kappan*. 63:106-112.
1983 Recruitment, selection, and retention: the shape of the teaching force. *The Elementary School Journal* 83(4):469-487.

Schmid, Rex
1980 *Final Report: Model Manpower Information System for Educational Personnel.* Gainesville: University of Florida.

Shivers, Joseph
1989 *Hiring Shortage-Area and Non Shortage-Area Teachers at the Secondary School Level*. Doctoral dissertation, Harvard Graduate School of Education.

Schwarzweller, H. K., and T. A. Lyson
1978 Some plan to be teachers: determinants of career specification among rural youth in Norway, Germany, and the United States. *Sociology of Education* 41:29-43.

Shymansky, James A., and Bill G. Aldridge
1982 The teacher crisis in secondary school science and mathematics. *Educational Leadership* November:61-62.

Sietsma, John
1981 *Teacher Layoffs, Shortages in 1979 Small Compared to Total Employed*. Special Report. National Center for Education Statistics. Washington, D.C.: U.S. Department of Education.

South Carolina State Department of Education
1985a *Teacher Need by Type*. Columbia, South Carolina: Author.
1985b *Teacher Supply and Demand for South Carolina Public Schools—An Initial Analysis—1984-85*. Columbia, South Carolina: Author.

Stevenson, Harold W.
1987 The Asian advantage: the case of mathematics. *American Educator* Summer:26-32.

Stevenson, H. W., and K. Bartsch
in press An analysis of Japanese and American textbooks in mathematics. In R. Leetsma and H. Walberg, eds., *Japanese Education*. Greenwich, Connecticut: JAI Press, Inc.

Stevenson, H. W., S. Lee, and M. Lummis
1988 Children's Understanding of Mathematics: China and the United States. Submitted for publication.

Stevenson, H. W., J. W. Stigler, G. W. Lucker, S. Lee, C. C. Hsu, and S. Kitemura
1986 Classroom behavior and achievement of Japanese, Chinese and American children. In R. Glaser, ed. *Advances in Instructional Psychology* 3:153-204. Hillsdale, New Jersey: Erlbaum.

Stigler, J. W., S. Lee, and H. W. Stevenson
1987 Mathematics classrooms in Japan, Taiwan, and the United States. *Child Development* 58:1272-1285.

Stiglmeier, J. J., and G. Rush
1982 *An Information Exchange*. Third edition. National Center for Education Statistics. Washington, D.C.: U.S. Department of Education.

Sweet, James A., and Linda A. Jacobsen
1983 Demographic aspects of the supply and demand for teachers. In Gary Sykes and Lee Shulman, eds., *Handbook of Teaching and Policy*. New York: Longmans.

Sykes, G.
1983 Contradictions, ironies, promises unfulfilled: a contemporary account of the status of teaching. *Phi Delta Kappan* 65:87-93.

Taylor, Suzanne S.
1986 *Public Employee Retirement Systems: The Structure and Politics of Teacher Pensions*. Ithaca, New York: Cornell University.

Tennessee State Higher Education Commission
1982 *A Study of Teacher Education in Tennessee*. Nashville, Tennessee: Author.

Thomas, R. B.
1975 The supply of graduates to school training. *British Journal of Industrial Relations* 13(March):107-114.

Travers, K. J.
1986 *Second Study of Mathematics, Detailed National Report—United States*. Champaign, Illinois: Stipes Publishing Company.

U.S. Department of Education
1989 Statistics reported by telephone, August 8. Office of the Mathematics and Science Education Program, U.S. Department of Education.

U.S. General Accounting Office
1984 *New Directions for Federal Programs to Aid Mathematics and Science Teaching*. Washington, D.C.: Author.

Vance, Victor S., and Phillip C. Schlechty
1982a The distribution of academic ability in the teaching force: policy implications. *Phi Delta Kappan* 64:22-27.
1982b The Structure of the Teaching Occupation and the Characteristics of Teachers: A Sociological Interpretation. Paper presented at the National Institute of Education Conference, Airlie House, Virginia.

Washington State
1984 *Professional Education: Annual Report. Special Theme: Teacher Supply and Demand 1983-84*. Olympia, Washington: Author.

Weaver, W.T.
1978 Educators in supply and demand: effects on quality. *School Review* 86:552-593.
1983 *America's Teacher Quality Problem: Alternatives for Reform*. New York: Praeger.

Weiss, Iris R.
1978 *Report of the 1977 National Survey of Science, Mathematics, and Social Studies Education*. Prepared for the National Science Foundation. Research Triangle Park, North Carolina: Research Triangle Institute.
1987 *Report of the 1985-86 National Survey of Science and Mathematics Education*. November. Research Triangle Park, North Carolina: Research Triangle Institute.

Westat, Inc.
1989 *NSF Teacher Transcript Study*. Prepared for the National Science Foundation. February. Rockville, Maryland: Westat, Inc.

Western Interstate Commission for Higher Education (WICHE)
1988 *High School Graduates: Projections by State, 1986 to 2004*. March. In cooperation with Teachers Insurance and Annuity Association and the College Board. Boulder, Colorado: Author.

Whitaker, Charles
1989 The disappearing black teacher. *Ebony* January.

Wimpelberg, R.K., and J. A. King
1984 Rethinking teacher recruitment. *Journal of Teacher Education* 34:5-8.

Wise, Arthur E., Linda Darling-Hammond, and Barnett Berry
1987 *Effective Teacher Selection: From Recruitment to Retention*. Santa Monica, California: The RAND Corporation.

Wood, K. E.
1978 What motivates students to teach? *Journal of Teacher Education* 29:48-50.

Wood, R. Craig
1982 The early retirement concept and a fiscal assessment model for public school districts. *Journal of Education Finance* 7(3)(Winter):262-276.

Wright, B. P.
1977 Our reason for teaching. *Theory Into Practice* 16:225-230.

Yamashita, June M.
1987 The Professional Development of Outstanding Mathematics Teachers. Pp. 65-69 in National Research Council, *The Teacher of Mathematics: Issues for Today and Tomorrow*. Proceedings of a Conference. Washington, D.C.: National Academy Press.

Zabalza, Antonio
1979 The determinants of teacher supply. *Review of Economic Studies* 46(January): 131-147.

Zarkin, G.
1985 Occupational choice: an application to the market for public school teachers. *Quarterly Journal of Economics* 100(2):409-446.

Appendix A
Panel Activities

Throughout the report we have referred to findings from our direct contacts with public school districts. To involve school district participation, the panel undertook three activities: a conference with personnel directors from large city school districts, a set of mini case studies building on reports of research in school districts studied by other researchers, and on-site in-depth case studies of school districts. Since the reader may want to evaluate the scope of these activities, they are described below.

CONFERENCE OF PERSONNEL DIRECTORS

A conference of personnel directors from seven large city school districts was convened in May 1988. The subject of this conference, "Structuring Professional Personnel Information Systems for Analyses of Teacher Supply and Demand," focused on supply- and demand-related data that large school districts regularly collect. Personnel administrators represented the following school districts:

- Seattle Public Schools (Washington)
- Montgomery County Public Schools (Maryland)*
- San Diego City Unified School District (California)
- Dade County Public Schools (Florida)
- Chicago Public Schools (Illinois)
- Los Angeles Unified School District (California)
- New York City Public Schools (New York)

* Also included in the mini case studies.

During the evening session on the first day of the conference the participants developed the following list of topics, which were discussed the next day:

1. *Effective recruiting strategies*

   What recruiting strategies are effective in attracting good teachers to urban districts with limited salaries?

   To what extent is it necessary to go outside the district to recruit?

   Are salary supplements for M/S teachers effective in recruitment?

   What problems arise from general recruiting rather than recruiting to meet need for each subject?

   How can general recruiting be used effectively to provide an adequate supply of M/S teachers?

   How can more people be attracted for M/S openings so district has a choice?

   What long-range effects on supply can be anticipated by aggressive recruiting to obtain a panel of applicants for each M/S position?

   Can the panel make recommendations that would affect recruitment problems?

2. *Experience with the reserve pool*

   What proportion of M/S teachers come from reserve pool?

3. Recruitment during the school year

   Why is it happening?

   Is it widespread among large and small districts?

   What are the reasons for vacancies during the school year?

   Are there trends in these reasons?

   What are the reasons for vacancies during the school year?

   Are there trends in these reasons?

   What effect does such recruiting have on teacher quality?

4. *Innovative approaches to address projected shortages of M/S teachers*

   Alternative certification programs

5. *Models used by districts for projecting need for teachers*

   What do districts actually do to project need for M/S teachers?

   Are projections limited by fact that available data were collected for administrative purposes such as hiring, paying, staffing, school buildings, and maintaining records for retirement?

   Is there a model for simulating staffing demands by subject that takes into account seniority rights to vacancies so that early hiring can be done in districts?

6. *Design of information systems that have the capability of identifying need for M/S teachers by subject*

   Need for integrated system

What knowledge should the information system be able to produce?

7. *Ideal information system*

   If there were no constraints on the information system, what information would you like to have?

8. *Teacher quality*

   How do districts define quality?

   Certification versus teaching out-of-field

   What information is helpful in recruiting for quality?

   Are elementary science teachers required to have science training, e.g., a laboratory science course?

   What effect have the NSF training institutes had on the quality of M/S teachers?

9. *Retraining*

   Has your district retrained teachers in fields such as social sciences to teach M/S subjects?

   Have retrained teachers been successful?

10. *Minorities and women*

    How can M/S teaching staff be balanced to provide role models for minorities and women?

    Are there enough women and minority M/S teachers to achieve such balance?

    What is needed to attract, train, and retain such teachers?

11. *Poor performance*

    What are the underlying issues in the relatively poor attainment of U.S. students in international math and science assessments?

    What could the NSF do to study these issues?

MINI CASE STUDIES

Only a small amount of information was collected in the mini case studies since the panel could draw on information about those districts reported by researchers who had conducted studies of these districts. A total of 27 mini case studies were conducted through a combined telephone-and-mail survey project in the summer and fall of 1988. The mini case studies involved the following districts:

- Houston Independent School District (Houston, Texas)
- Hillsborough County Public Schools (Tampa, Florida)
- Montgomery County Public Schools (Montgomery County, Maryland)
- Clark County School District (Las Vegas, Nevada)
- Jefferson County School District (Louisville, Kentucky)

- New Orleans Public Schools (New Orleans, Louisiana)
- Albuquerque Public Schools (Albuquerque, New Mexico)
- Charlotte-Mechlenburg Schools (Charlotte, North Carolina)
- Columbus Public Schools (Columbus, Ohio)
- Austin Independent School District (Austin, Texas)

- Mesa Unified School District (Mesa, Arizona)
- Rochester City School District (Rochester, New York)
- Richland School District No. 1 (Columbia, South Carolina)
- Salt Lake City School District (Salt Lake City, Utah)
- Guilford County Schools (Greensboro, North Carolina)

- Lake Washington School District (Kirkland, Washington)
- Durham County Schools (Durham, North Carolina)
- Greenwich Public Schools (Greenwich, Connecticut)
- Barrow County School District (Winder, Georgia)
- Martin County Public Schools (Williamston, North Carolina)

- Watauga County Schools (Boone, North Carolina)
- Northampton County Schools (Jackson, North Carolina)
- Jamestown Public Schools (Jamestown, North Dakota)
- Howard-Suamico School District (Green Bay, Wisconsin)
- MSAD NO. 15 (Gray-New Gloucester, Maine)

- East Williston Unified School District (East Williston, New York)
- Medicine Valley School District (Curtis, Nebraska)

The telephone interview guide and the mail-questionnaire form that were used for the 27 mini case studies can be found at the end of this appendix.

IN-DEPTH CASE STUDIES

Six in-depth case studies were conducted on site, involving school districts in California, Maryland, and Utah. The school districts are not named because confidentiality was pledged.

Jane L. David conducted case studies of two neighboring districts in California that were expected to draw on the same labor market. Marianne Amarel conducted case studies of a pair of adjacent districts in Maryland. Finally, two additional school districts were selected—one in Utah and one in California—because they were experiencing substantial increases in enrollment. Special problems of supply and demand for science and mathematics teachers may exist in districts with increasing enrollment. And since secondary enrollments are projected to increase nationwide in the near future, we wanted to include in-depth studies of districts now experiencing increases. Jane L. David and Jennifer Pruyn conducted these case studies.

(Telephone Interview Guide for Mini Case Studies)

HS GRADE:

DISTRICT:

INTERVIEWEE:

BACKGROUND:

1. During which months do you interview/hire?

2. Looking for m/s teachers with special qualifications?

3. Particular difficulty recruiting m/s teachers?

4. Does m/s recruiting differ from recruiting for other subjects?

5. Categories of m/s teachers in your district records:

(Survey Questionnaire for Mini Case Studies)

Supply and Demand for
High School Mathematics and Science Teachers

_______________ District Name
_______________ Person Completing Form
_______________ Position Title
_______________ Telephone Number

District Context

1. What is your district's total enrollment? _______________

2. What is the current approximate ethnic mix of your district's student body? (____ % white, ____ % black, ____ % Spanish surname, ____ % other)

3. Has your district experienced any reductions-in-force during the period 1985-1988? _______________

4. During the period 1985-1988, was the high school student population growing, stable, or decreasing? _______________

5. Number of high schools in your district: _______________

High School Mathematics and Science Teachers

6. How many high school mathematics and science (m/s) teachers are currently employed by your district? _______________

7. How many high school m/s teachers have 5 or more years of service in your district? _______________ 10 or more years? _______________

8. What is the starting salary for a m/s teacher with a BA and no prior experience? _______________
   What is the top salary for a m/s teacher with a master's degree? _______________

9. Where do most of your m/s applicants come from? (i.e., nearby universities or teacher training institutions, other districts, etc.) _______________

10. About how many fully qualified applicants per vacancy do you have in mathematics? _______________ In science? _______________

11. Is your district experiencing shortages of qualified applicants in mathematics or science subjects? ______ If so, in what subjects? _______________

12. How many vacancies were filled for all high school subjects, and in particular for mathematics and science?

	Total (all subjects)	Mathematics	Science
This school year (1987-1988)	____	____	____
Last school year (1986-1987)	____	____	____
The school year before (1985-1986)	____	____	____

13. What was the principal reason for the m/s vacancies in these 3 years?

____ To respond to enrollment growth
____ To replace retirees
____ To replace teachers leaving the system for reasons other than retirement

Comment: ____
____

14. Over the past 3 school years, how many of your m/s vacancies were filled by persons with the following kinds of experience/certification in teaching m/s?

	This school year (1987-88)	Last school year (1986-87)	The school year before (1985-86)
Certified in m/s:			
o New graduates with no prior teaching experience	____	____	____
o Earlier graduates but with no prior teaching experience	____	____	____
o Experienced m/s teachers	____	____	____
Noncertified in m/s, but with emergency or temporary credentials	____	____	____

15. Does the district sponsor training programs for teachers who are not certified in m/s, but who are filling m/s vacancies? ____ Comment: ____
____
____

Appendix B
National Data Sets Relating to Demand, Supply, and Quality of Precollege Science and Mathematics Teachers

The Schools and Staffing Survey (SASS)
High School and Beyond
The National Longitudinal Study (NLS-72)
National Education Longitudinal Study of 1988 (NELS:88)
Surveys of Recent College Graduates (RCG)
The National Assessment of Educational Progress (NAEP)
The American Freshman
National Surveys of Science and Mathematics Education
Status of the American Public School Teacher

THE SCHOOLS AND STAFFING SURVEY

The Schools and Staffing Survey (SASS) is an integrated set of the National Center for Education Statistics surveys consisting of the Teacher Demand and Shortage Survey, the School Survey, the School Administrator Survey, and the Teacher Survey (including separate follow-ups a year later of those sampled teachers who leave teaching and a subsample of those who stay in teaching). These surveys were first conducted in the 1987-88 school year, will be conducted again in 1990-91, and are scheduled to be repeated thereafter every two years. They are designed to better measure important aspects of teacher supply and demand, the composition of the administrator and teacher work force, and the status of teaching and schooling generally. More specifically, five purposes underlie these studies: (1) to profile the nation's precollege teaching force; (2) to improve estimates and projections of teacher supply and demand by teaching field, sector, level, and geographic location; (3) to allow analyses of teacher mobility and turnover;

(4) to enhance assessment of teacher quality and qualifications; and (5) to provide more complete information on school policies and programs, administrator characteristics, and working conditions. If implemented successfully, we will have a national data base of indicators of teacher supply, demand, and quality.

The sampling unit for SASS is the school—9,300 public and 3,500 private schools were selected for SASS; the districts to which those schools belonged were then identified for the sample. Thus, 5,600 public local education agencies are in the sample (of a universe of 16,000). Within the selected schools, 52,000 public and 13,000 private school teachers were sampled, totaling 65,000 teachers. Of the 52,000 public school teachers, 23,000 taught at the secondary level; 29,000 were elementary teachers.

Below are capsules of the kinds of data found in each of the surveys.

SASS 1A—Teacher Demand and Shortage Questionnaire for Public School Districts (SASS 1B is the parallel private school form). District enrollment, hiring and retirement policies, and staff data. Number of teaching positions, by level and field, that are filled or remain unfilled. New hires, layoffs, salaries, benefits. High school graduation requirements by field.

SASS 2—School Administrator Questionnaire (public and private). Training, experience, and professional background of principals. School problems, including teacher absenteeism. Influence of teachers/principal/district on curriculum and on hiring. Methods of dealing with unfilled vacancies.

SASS 3A—Public School Questionnaire (SASS 3B is the parallel private school form). Pupil-teacher ratio, student characteristics, staffing patterns, and teacher turnover (entry, attrition). Supply sources of new entrants and destinations of leavers. Some data can be analyzed by academic subject area.

SASS 4A—Public School Teacher Questionnaire (SASS 4B is the parallel private school form). Education and training, current assignment, continuing education, job mobility, working conditions, career choices. Division of time, courses taught. Achievement level of students. Salary, other income. Opinions on pay policies, salary, working conditions, professional recognition, etc. What they did before they began teaching at this school. Data can be analyzed by teaching field.

TFS 2—Teacher Follow-up Survey (Questionnaire for former teachers sampled in SASS). Teacher attrition, salary, other factors or reasons for leaving teaching. What they did after leaving. Comparison of teaching with current occupation with regard to salary, working conditions, and job satisfaction. Data can be analyzed by subject.

TFS 3—Teacher Follow-up Survey (Questionnaire for sampled teachers who remained in teaching). Factors in retention; reasons for possible change in school assignment. Salary, other income. Data can be analyzed by subject.

Data related to demand are found in SASS 1A, the district-level questionnaire; data related to supply, teacher qualifications, and quality are found in the teacher questionnaires. Links exist among the school, teacher, and administrator questionnaires to enable comparative analyses. And repetition of SASS every two years will yield valuable information on trends in indicators over time.

Contact: Mary Papageorgiou
National Center for Education Statistics
555 New Jersey Avenue, N.W.
Washington, D.C. 20208
202/357-6336

HIGH SCHOOL AND BEYOND

High School and Beyond is a national longitudinal survey of 1980 high school seniors and sophomores conducted by the National Center for Education Statistics. A probability sample of 1,015 public and private high schools was selected with 36 seniors and 36 sophomores in each of the schools. A total of about 30,000 sophomores and 28,000 seniors participated in the base-year survey. The base-year data were collected in 1980, with follow-ups in 1982, 1984, and 1986. In addition, data from their parents and teachers and high school and postsecondary education transcripts were included. Currently the sample contains 14,825 participants from the 1980 sophomore cohort and 11,995 participants from the 1980 senior cohort.

The purpose of the survey is to observe the educational, occupational, and family development of young people as they pass through high school and college and take on adult roles. Data obtained can also help researchers understand the new graduate component of the supply pool and the incentives to which they respond.

The 1980 and 1982 surveys consisted of questionnaire data (on background characteristics, attitudes, postsecondary educational and career plans, and activities related to education, career, and family development). Cognitive tests developed for the sophomore cohort by the Educational Testing Service were administered in 1980 and 1982. The tests were designed to measure cognitive growth in three domains: verbal, mathematics, and science.

The 1984 and 1986 follow-up surveys contain similarly detailed questions concerning college courses and experiences, jobs (including salaries), attitudes, and marriage and family formation. Tentative plans call for an additional follow-up of the 1980 sophomores in 1991. An analysis file is being prepared containing transcripts and student responses for students indicating that they plan to become teachers.

The Administrator and Teacher Survey (ATS) was designed and given to a sample of High School and Beyond school staff in 1984 to explore findings from "effective schools" research with a broadly representative sample. The effective schools literature identifies characteristics of schools in which students perform at higher levels than would be expected from their background and other factors. Prior to the ATS, measures of those characteristics were not available on any large national data set. The ATS provides measures of staff goals, school environment, school leadership, and other processes believed important.

A total of 457 public and private high schools (approximately half of the 1,015 High School and Beyond schools) were sampled for the ATS; separate questionnaires were prepared for principals, teachers, vocational education coordinators, heads of guidance, and community service coordinators. Up to 30 teachers in each of the 457 schools responded to the teacher questionnaire; only one respondent per school completed the other surveys. There are 402 principals in the sample. In all, approximately 11,000 administrators and teachers participated.

The ATS was designed to measure school goals and processes that the effective schools literature indicates are important in achieving effective education. Questionnaire items describe staff goals, pedagogic practices, interpersonal staff relations, teacher workload, staff attitudes, availability and use of services, planning processes, hiring practices, optional programs designed to produce educational excellence, and linkage to local employers, parents, and the community.

The ATS asked teachers a number of quality-related questions concerning school environment, in-service experience, interruptions, autonomy, absenteeism, parent contact, hours spent teaching and nonteaching, and time use and practices in a typical class. The respondent's educational background and subject preparation, certification and salary data were also asked. NCES has not issued publications based on the ATS, but the data are available on tape and a code book is available.

Contact: Carl Schmitt
National Center for Education Statistics
555 New Jersey Avenue, N.W.
Washington, D.C. 20208
202/357-6772

THE NATIONAL LONGITUDINAL STUDY

The National Longitudinal Study (NLS-72), conducted by NCES and administered by the National Opinion Research Center, studies the high school class of 1972 in the form of a sample of 23,000 high school seniors (1972) enrolled in 1,318 high schools. Follow-ups were conducted in 1973-74, 1974-75, 1976-77, and 1979-80. A fifth follow-up was conducted on a subsample in 1986, with approximately 13,000 responding; at this point they were about 32 years old. In each follow-up, data were collected on high school experiences, background, opinions and attitudes, and future plans. Participants took achievement tests in the first survey. Follow-ups traced their college, postgraduate and work experiences, including salaries. Reasons for leaving schools or jobs were also asked. In addition, respondents included data on marriage and family formation and military service.

A Teaching Supplement Questionnaire was sent to all respondents to the fifth follow-up survey (1986) who indicated they had teaching experience or had been trained for precollege teaching. In addition, persons with mathematics, science or engineering backgrounds (with 2-year, 4-year, or graduate degrees in those fields) were drawn into the sample. A total of 1,147 eligible individuals responded. Of these, 109 indicated they actually had no teaching experience, degree in education, or certification to teach. This left 1,038 completed teaching supplements to analyze, drawing on the wealth of previous NLS data on these individuals.

This sample of current and former teachers (and some who never became teachers) were asked about career paths, salaries in teaching and nonteaching positions, certification, continuing education, family formation, reasons for entry into teaching and attrition, and nonteaching jobs. This detailed information can be analyzed by subject area.

The data have been analyzed at NCES and by Heyns (1988) in light of contributing to knowledge of the characteristics of the supply pool, to identifying patterns of entry, exit, and reentry and to understanding the role of salary and other incentives.

Contact: Paula Knepper
National Center for Education Statistics
555 New Jersey Avenue, N.W.
Washington, D.C. 20208
202/357-6914

NATIONAL EDUCATION LONGITUDINAL STUDY OF 1988

The National Education Longitudinal Study of 1988 (NELS:88) is a new education longitudinal study sponsored by NCES and designed to provide trend data about critical transitions experienced by young people as they develop, attend school, and embark on their careers. By initially focusing in 1988 on 8th graders and their schools, teachers, and parents, then by following up that cohort at two-year intervals, the NELS:88 data will be used to address such issues as persistence and dropping out of high school, transition from 8th grade to high school, tracking, and features of effective schools.

For the base-year survey conducted in the spring of 1988, a nationally representative sample of 1,000 schools (800 public and 200 private) was drawn. Within this school sample, 26,000 8th grade students, 6,000 8th grade teachers, and 24,000 parents were surveyed. Thus, the four major component surveys for the base year were directed at students, parents, school administrators, and teachers.

Students were asked about school work, aspirations, and social relations. They also took cognitive tests in four achievement areas: reading and vocabulary, mathematics, science, and social studies. The parent survey gauged parental aspirations for their children, commitment of resources to their children's education, and other family characteristics relevant to educational achievement. Analysis of these data may suggest young people's levels of interest and their parents' commitment to pursuing science/mathematics fields in the future.

School principals provided information about the teaching staff, student body, school policies and offerings, and courses required for 8th graders. For example, for science and for mathematics, the principal was to note whether a full year, a half year, less than a half year, or no specified amount is required. Whether a gifted-talented program is offered is also noted, by subject. School environment items, and particularly discipline indicators, are included. Staffing questions are general and not broken down by subject. From this survey, data on 8th grade mathematics or science required might inform demand in a general way. Course-taking data might inform preparation for high school mathematics and science. Indicators of quality of education offered at the 8th grade level are somewhat more evident.

Teacher data include academic background and certification information, class size, time use, instructional materials used, laboratory use, and school environment information. The teacher questionnaire thus can shed light on a number of indicators of quality of education offered at the 8th grade level by subject.

The science and mathematics teachers who participated in NELS:88 are expected to furnish postsecondary education transcripts for a National Science Foundation study, the NSF Teacher Transcript Study, begun under contract with Westat, Inc., in 1988. This study plans to collect transcript data on a national basis for use in assessing teacher characteristics.

Contact: Jeffrey Owings
National Center for Education Statistics
555 New Jersey Ave., N.W., Room 518
Washington, D.C. 20208
202/357-6777

SURVEYS OF RECENT COLLEGE GRADUATES

The National Center for Education Statistics has conducted periodic surveys, as described above, on outcomes of college graduation. The Recent College Graduates (RCG) surveys, which are not longitudinal, have concentrated especially on those graduates entering the teaching profession. Education majors are thus oversampled for the RCG. These surveys have primarily addressed the issues of employment related to individuals' field of study and their access to graduate or professional programs.

The survey involves a two-stage sampling procedure. First, a sample of institutions awarding bachelor's or master's degrees is selected and stratified by percent of education graduates, control, and type. Special emphasis is placed on institutions granting degrees in education and on traditionally black institutions. For each of the selected schools, a sample of degree recipients is chosen. Included are both B.A. and M.A. degree recipients.

The survey of 1974-75 college graduates was the first and smallest of the series. The sample consisted of 200 responding schools. Of the 5,506 graduates in that sample, 4,350 responded (79 percent). The 1981 survey was somewhat larger, covering 301 institutions and 15,852 students. The student response rate was 62 percent. The 1985 survey (which collected race/ethnicity data for the first time) requested data from 18,738 students from 404 colleges. The student response rate was an effective 70 percent, with just under 11,000 participating. Response rates in these cycles (except for the 1976 survey) tend not to be higher because of invalid mailing addresses, reflecting the difficulty in tracing students after graduation. The 1987 study (which included transcripts for the first time) was more effective in locating graduates as the file contains 16,878 respondents from 400 higher education institutions, representing an 80 percent response rate. This study was also the largest sample (21,957 eligible sample members) drawn from institutions to date. The RCG may thus be biased against more mobile graduates. Students are surveyed once, without additional follow-ups.

Questionnaire items request data including degrees and teaching certificates; continuing education; additional formal training; what job was held as of April 27, 1987, and its relation to educational training; subjects eligible/certified to teach, whether/when entered teaching; subjects taught; marital status and number of children; and further degree plans. The questionnaire asks whether the person taught in grades K-12 before completing the degree requirement. Subjects taught are phrased generally: mathematics, computer science, biological science, and physical sciences. Information on the graduates' incentives for choosing particular careers or jobs is limited.

Thus, the RCG offers a general, nonlongitudinal picture of the new graduates component of the supply pool. Its inclusion of items that may suggest incentives and disincentives to enter teaching make it a possible source of reserve pool information. In addition to supply related to potential teachers and potential minority teachers, a few aspects of the qualification component of quality may be touched on rather indirectly, such as grades (self-reported).

The RCG will be conducted again in 1991. Beginning in 1994, the RCG is scheduled to be redesigned as a longitudinal study, the Baccalaureate and Beyond Longitudinal Study.

Contact: Martin M. Frankel
National Center for Education Statistics
555 New Jersey Avenue, N.W.
Washington, D.C. 20208
202/357-6774

THE NATIONAL ASSESSMENT OF EDUCATIONAL PROGRESS

The National Assessment of Educational Progress (NAEP) is a congressionally mandated study directed and funded by the National Center for Education Statistics. The assessment is currently administered for NCES by the Educational Testing Service. It is referred to as The Nation's Report Card, the National Assessment of Educational Progress. Approximately 120,000 precollege students are randomly selected for the national assessment every two years. The overall goal is to determine the nation's progress in educational achievement, including achievement in science and mathematics.

To accomplish this goal, NAEP has surveyed the educational accomplishments of 9-, 13-, and 17-year-old students in 11 subject areas, starting in 1969-70. NAEP first identifies counties as primary sampling units through a stratified sampling plan. Then for each age level, public and private schools are selected by a stratified sampling plan. Finally, within each school groups of students are selected to participate in NAEP.

The most recent science and mathematics assessments were in 1985-86; the next ones are planned for 1990. Previous mathematics assessments were in 1973, 1976, 1978, and 1982. The science assessment was previously given in 1969-70, 1973, 1977, and 1982. By law, mathematics is now to be assessed every two years and science every four years.

Assessments are given in fall, winter, and spring, measuring achievement and gathering information on attitudes and classroom practices as students perceive them. The nonachievement measures (attitudes toward science or math, homework and grades, and home environment) are obtained through a companion background questionnaire.

NAEP Teacher Questionnaire

In 1984, NAEP began collecting data on teacher attributes, as reported by teachers of the students participating in the NAEP assessments. At grades 7 and 11, teachers are identifiable by subject (e.g., mathematics, science). The 1986 assessment, for example, gathered information from 325 7th grade science teachers and 289 11th grade science teachers who responded.

The teacher questionnaire asked for data on general demographic characteristics, certification, educational preparation, and teaching experience at various grade levels. These help to illuminate aspects of teacher qualifications. School environment indicators asked on the questionnaire include classroom activities and practices, homework, laboratory and other instructional resources, and autonomy. The questionnaire asks whether the respondent would become a teacher if he or she could start over again. Continuing education is touched on in one item, although not in detail. There are no questions tracing the teacher's career path, salary, or other nonteaching work. Thus, the NAEP teacher survey may provide some quality-related information, but little on demand or supply.

In 1990, mathematics teachers in grades 4 and 8 will be surveyed. Science teachers at grade 8 will also receive questionnaires. The Horizon Corporation is under contract to develop the science teacher questionnaire for 1990.

Contact: Kent Ashworth
Educational Testing Service
Princeton, New Jersey 08541
1/800/223-0267

Gene Owen
National Center for Education Statistics
555 New Jersey Avenue, N.W.
Washington, D.C. 20208
202/357-6746

THE AMERICAN FRESHMAN

The American Freshman survey is conducted annually by the Cooperative Institutional Research Program (CIRP), of the University of California at Los Angeles (UCLA). CIRP and UCLA's Higher Education Research Institute survey all incoming freshmen in full-time study in a sample of colleges and universities. The data are stratified by type of college, public or private control, and selectivity. Longitudinal follow-up studies are conducted each summer to track students two and four years after college entry. Freshman surveys typically involve 300,000 students at 600 institutions; follow-ups are done with random samples of 25,000 students from each cohort.

The 40-question survey instrument solicits data on high school background, including SAT or ACT scores and grade point average, intended major and educational goals, career plans, financial arrangements, and attitudes. Personal data include race/ethnicity, sex, and parents' income and occupations. Data from The American Freshman can illuminate the beginning stage of the supply pipeline—choosing a major and a career plan. Questionnaire data, such as SAT scores and number of honors courses taken in high school, can be used to provide some measure of the qualifications aspects of quality.

Contact: Alexander W. Astin
Higher Education Research Institute
University of California
Los Angeles, California 90024
213/825-4321

NATIONAL SURVEYS OF SCIENCE AND MATHEMATICS EDUCATION

The 1977 and 1985 surveys were sponsored by the National Science Foundation and conducted by Iris Weiss of Research Triangle Institute. They involved a national probability sample of schools, principals, and teachers in grades K-12. The 1985 survey covered 425 public and private schools. From these schools a sample of 6,000 teachers was selected. The sample was stratified by grades K-6, 7-9, and 10-12. For grades 10-12 the sample was also stratified by subject to avoid the oversampling of biology. Of the teachers sampled, 2,300 were teaching at the grade 10-12 level. Response rates were generally high; for example, the response rate from principals was 86 percent.

Principals in schools selected for the grade 10-12 sample were asked to check the types and number of science and mathematics courses taught by each teacher: biology/life sciences, chemistry, physics, earth/space science, "other mathematics/computer science." Principals also reported whether they had difficulty hiring fully qualified teachers for vacancies, by subject.

The survey requested science and mathematics course offerings and enrollment (by race, ethnicity, and sex), science labs and equipment, instructional techniques, and teacher training. Information on achievement was not requested.

Teachers supplied in-depth information on curriculum and instruction in a single, randomly selected class. Time spent in instruction, lab, and amount of homework given are among the types of practices for which data were collected.

In addition, data on the detailed educational background of each teacher were requested. Information about advanced degrees earned, certification, and subject-matter courses taken, to compare with standards of the NSTA and NCTM. Teacher age and teaching experience were also included.

Most of the teacher data from the National Survey of Science and Mathematics Education are related to aspects of quality. From the school-level data supplied by principals, indicators of demand may be found to some extent in the course offerings and enrollment data.

Contact: Iris Weiss or Jennifer McNeill
Research Triangle Institute
P.O. Box 12194
Research Triangle Park, North Carolina 27709
1/800/334-8571

STATUS OF THE AMERICAN PUBLIC SCHOOL TEACHER

National Education Association (NEA) in 1956 developed the first of a series of surveys covering numerous aspects of U.S. public school teachers' professional, family, and civic lives. This survey project, titled The Status of the American Public School Teacher, has been conducted every five years since 1956. Although the questionnaire has been revised to update items of concern, the wording still provides comparable data on most items from survey to survey (except for 1961, which contained some differences in the wording of questions). The most recent survey was conducted in 1985-86 and published by the NEA in 1987.

The sample of respondents for 1985-86 contained 1,291 usable responses (72.4 percent of the questionnaires originally mailed). Participants were selected through a two-stage sample design: first, a sample of public school districts was drawn, classified by pupil enrollment into nine strata. All school districts in the sample were asked to submit a list of all their teachers. Using that list, systematic sampling with a random start was used. A 58-item questionnaire was then mailed, in spring 1986, to all teachers in the sample. Questionnaire items span teaching experience, educational background, subject(s) taught, income, workload, school environment, demographic and family information, and civic interests. Subject area taught is self-reported, with the teacher filling in a blank with the main subject taught (i.e., "science").

Items related to supply include number of breaks in service and (one) primary reason, salaries from teaching and from additional employment, what the person did the previous year, what he or she plans to do next year, and how long the person plans to remain in teaching.

Items related to quality include highest college degree and recentness of that degree; teaching and nonteaching loads; type of teaching certificate held; college credits earned in the past three years and how much of the teacher's own money was spent for credits and other school expenses; detailed information about professional growth activities (workshops, university extension, college courses in education/other than education, etc.); whether the person would become a teacher if he or she started over again; reasons for teaching; what helps/hinders the teacher most in his or her position; and presence of teaching assistants.

Contact: Richard R. Verdugo
Research Division
National Education Association
1201 Sixteenth St., N.W.
Washington, D.C. 20036
202/822-7400

Appendix C
Availability of State Data on Public School Professional Personnel

Availability of State Data on Public School Professional Personnel

	File characteristics by state										
	AL	AK	AZ	AR	CA	CO	CT	DE	DC	FL	GA
Is an automated file maintained?	yes	yes	yes	yes	yes	----	yes	yes	----	no	yes
Earliest year available:	---	84	81	83	81	---	78	81	---	---	85
How often is the file updated? (daily, monthly, annually)	D	A	A	A	A	---	A	A	---	---	e
Data elements in files											
School, personal employment data:											
School district code	yes	yes	yes	yes	yes	---	yes	yes	---	---	yes
School code	yes	yes	---	yes	yes	---	yes	yes	---	---	yes
Social Security number	yes	yes	yes	yes	yes	---	yes	yes	---	---	yes
Date of birth/age	yes	yes	yes	yes	yes	---	yes	yes	---	---	yes
Sex	yes	yes	yes	yes	yes	---	yes	yes	---	---	yes

Availability of State Data on Public School Professional Personnel--continued

	File characteristics by state										
	AL	AK	AZ	AR	CA	CO	CT	DE	DC	FL	GA
Racial/ethnic group	yes	yes	yes	yes	yes	---	yes	yes	---	---	---
Ed. level/degree status	yes	yes	yes	yes	yes	---	yes	yes	---	---	---
Acad. major (bachelor's)	yes	yes	---	yes	---	---	---	---	---	---	---
Annual contract salary	yes	yes	yes	yes	yes	---	yes	yes	---	---	f
Type of appointment	yes	---	---	---	yes	---	yes	yes	---	---	g
Number of months employed	yes	---	---	yes	---	---	---	yes	---	---	---
Percentage of time employed	yes	yes	yes	yes	yes	---	yes	yes	---	---	yes
Information on assignments	yes	yes	yes	yes	yes	---	yes	yes	---	---	---

Teaching experience:											
Total years	yes	---	---	yes	yes	---	c	yes[d]	---	---	yes
Years in current district	yes	---	---	---	yes	---	---	---	---	---	---
Years in other district(s)	yes	---	---	---	---	---	---	---	---	---	---
Years in current assignment	---	---	---	---	yes	---	---	---	---	---	---
For new hires:											
Occupation prior year	---	---	---	---	---	---	yes	yes	---	---	---
Location of occupation prior year	---	---	---	---	---	---	---	yes	---	---	---
Other data:	---	---	---	---	b	---	---	---	---	---	h
Courses/levels taught	yes	yes	yes	---	yes	---	yes	yes	---	---	---

Availability of State Data on Public School Professional Personnel--continued

	File characteristics by state									
	HA	ID	IL	IN	IA	KS	KY	LA	ME	MD
Is an automated file maintained?	yes	yes	yes	yes	yes	yes	yes	yes	yes	yes
Earliest year available:	74	---	63	62	73	80	70	87	70	85
How often is the file updated? (daily, monthly, annually)	D	A	A	A	A	A	A	A	A	A
Data elements in files										
School, personal employment data:										
School district code	yes	yes	yes	yes	yes	yes	yes	yes	yes	yes
School code	yes	yes	yes	yes	yes	yes	yes	yes	yes	yes
Social Security number	yes	yes	yes	yes	yes	yes	yes	yes	yes	yes
Date of birth/age	yes	yes	yes	yes	yes	yes	yes	---	yes	yes
Sex	yes	yes	yes	yes	yes	yes	yes	---	yes	yes

Racial/ethnic group	yes	---	yes	yes	---	yes	yes	---	yes	yes
Ed. level/degree status	yes	yes	yes	yes	yes	yes	yes	yes	yes	yes
Acad. major (bachelor's)	yes	yes	---	---	yes	---	yes	yes	yes	---
Annual contract salary	yes	yes	yes	yes	yes	yes	yes	---	yes	yes
Type of appointment	yes	---	---	---	---	---	---	---	---	---
Number of months employed	---	yes	yes	yes	yes	---	yes	---	---	yes
Percentage of time employed	yes	---	yes	yes	yes	yes	yes	---	yes	yes
Information on assignments	---	yes	yes	yes	yes	yes	yes	yes	yes	yes

Availability of State Data on Public School Professional Personnel--continued

	File characteristics by state									
	HA	ID	IL	IN	IA	KS	KY	LA	ME	MD
Teaching experience:										
Total years	yes	yes	yes	yes	yes	yes	yes	---	yes	yes
Years in current district	---	---	yes	yes	yes	yes	---	---	yes	---
Years in other district(s)	---	---	yes	---	---	yes	---	---	yes	---
Years in current assignment	---	yes	---	---	---	---	---	---	---	---
For new hires:										
Occupation prior year	---	---	---	yes	---	yes	yes	---	yes	---
Location of occupation prior year	---	---	---	---	---	---	---	---	---	---
Other data:	---	---	---	i	---	---	---	---	j	k
Courses/levels taught	---	---	---	yes	yes	yes	yes	yes	yes	yes

Availability of State Data on Public School Professional Personnel--continued

	File characteristics by state									
	MA	MI	MN	MS	MO	MT	NE	NV	NH	NJ
Is an automated file maintained?	no	yes	yes	yes	yes	yes	yes	yes	yes	yes
Earliest year available:	---	61	---	81	82	---	80	86	87	77
How often is the file updated? (daily, monthly, annually)	---	A	A	A	A	A	A	A	A	A
Data elements in files										
School, personal employment data:										
School district code	---	yes	yes	yes	yes	yes	yes	yes	yes	yes
School code	---	yes	yes	yes	yes	yes	yes	yes	yes	yes
Social Security number	---	yes	---	yes	yes	yes	yes	yes	yes	yes
Date of birth/age	---	yes	yes	yes	---	---	yes	yes	yes	yes
Sex	---	yes	yes	yes	yes	yes	yes	yes	yes	yes

Availability of State Data on Public School Professional Personnel--continued

	File characteristics by state									
	MA	MI	MN	MS	MO	MT	NE	NV	NH	NJ
Racial/ethnic group	---	yes	---	yes	---	yes	yes	yes	---	yes
Ed. level/degree status	---	yes	yes	yes	yes	---	yes	yes	yes	yes
Acad. major (bachelor's)	---	yes	yes	yes	---	---	---	yes	yes	yes
Annual contract salary	---	---	yes	yes	yes	yes	yes	yes	---	yes
Type of appointment	---	---	---	yes	---	---	---	---	---	---
Number of months employed	---	---	yes	yes	---	yes	---	---	---	yes
Percentage of time employed	---	---	---	yes	yes	yes	yes	---	---	yes
Information on assignments	---	yes	yes	yes	yes	yes	yes	yes	yes	yes

Teaching experience:										
Total years	---	yes	yes	yes	yes	---	yes	---	yes	yes
Years in current district	---	---	---	---	yes	yes	yes	---	yes	yes
Years in other district(s)	---	---	---	---	---	---	---	---	yes	yes
Years in current assignment	---	yes	---	---	---	---	---	---	yes	yes
For new hires:										
Occupation prior year	---	---	---	---	---	---	---	---	---	yes
Location of occupation prior year	---	yes	---	---	---	---	---	---	---	---
Other data:	l	---	m	---	---	---	---	---	n	o
Courses/levels taught:	yes	yes	yes	yes	yes	yes	yes	yes	yes	yes

Availability of State Data on Public School Professional Personnel--continued

	File characteristics by state									
	NM	NY	NC	ND	OH	OK	OR	PA	RI	SC
Is an automated file maintained?	yes	yes	yes	yes	yes	yes	yes	yes	yes	yes
Earliest year available:	87	67	78	---	71	70	70	70	78	81
How often is the file updated? (daily, monthly, annually)	---	A	A	A	A	D	A	A	D	A
Data elements in files										
School, personal employment data:										
School district code	yes	yes	yes	yes	yes	yes	yes	yes	yes	yes
School code	yes	yes	yes	yes	yes	yes	yes	yes	yes	yes
Social Security number	yes	yes	yes	yes	yes	---	yes	yes	yes	yes
Date of birth/age	yes	yes	yes	yes	yes	---	yes	yes	yes	yes
Sex	yes	yes	yes	yes	yes	yes	yes	yes	yes	yes

Racial/ethnic group	yes	---	yes	yes	yes	yes	---	yes	---	yes
Ed. level/degree status	yes	yes	yes	yes	yes	yes	yes	yes	yes	yes
Acad. major (bachelor's)	yes	---	---	yes	---	yes	---	---	yes	---
Annual contract salary	yes	yes	yes	yes	yes	yes	yes	yes	---	yes
Type of appointment	---	yes	---	---	---	---	---	---	---	---
Number of months employed	---	yes	yes	yes	---	yes	yes	---	---	yes
Percentage of time employed	yes	yes	yes	yes	---	yes	yes	yes	---	yes
Information on assignments	yes	yes	yes	yes	yes	yes	yes	yes	yes	yes

Availability of State Data on Public School Professional Personnel--continued

	File characteristics by state									
	NM	NY	NC	ND	OH	OK	OR	PA	RI	SC
Teaching experience:										
Total years	yes	yes	yes	yes	yes	yes	---	yes	yes	yes
Years in current district	yes	yes	yes	---	---	yes	yes	yes	---	yes
Years in other district(s)	yes	yes	yes	---	---	yes	yes[u]	---	---	yes
Years in current assignment	---	yes	yes	---	---	---	---	---	---	---
For new hires:										
Occupation prior year	---	yes	[q]	yes	---	---	---	---	---	---
Location of occupation prior year	---	yes	---	---	---	---	yes	---	---	---
Other data:	---	[p]	[r]	[s]	---	[t]	[u]	[v]	[w]	[y]
Courses/levels taught	yes	yes	yes	yes	yes	---	yes	yes	yes	yes

Availability of State Data on Public School Professional Personnel--continued

	File characteristics by state									
	SD	TN	TX	UT	VT	VA	WA	WV	WI	WY
Is an automated file maintained?	yes	yes	yes	yes	yes	yes	yes	yes	yes	yes
Earliest year available:	82	81	74	82	---	81	---	76	75	87
How often is the file updated? (daily, monthly, annually)	A	D	D	D	---	A	A	A	A	cc
Data elements in files										
School, personal employment data:										
School district code	yes	yes	yes	yes	yes	yes	yes	yes	yes	yes
School code	yes	---	yes	yes	yes	yes	yes	yes	yes	yes
Social Security number	yes	yes	yes	yes	yes	yes	yes	yes	yes	yes
Date of birth/age	yes	---	yes	yes	yes	yes	yes	yes	yes	yes
Sex	yes	yes	yes	yes	yes	yes	yes	yes	yes	yes

Availability of State Data on Public School Professional Personnel--continued

	File characteristics by state									
	SD	TN	TX	UT	VT	VA	WA	WV	WI	WY
Racial/ethnic group	---	yes	yes	yes	---	yes	yes	---	yes	yes
Ed. level/degree status	yes	yes	yes	yes	yes	yes	yes	yes	yes	yes
Acad. major (bachelor's)	yes	yes	yes	yes	---	---	---	---	---	yes
Annual contract salary	yes	yes	yes	yes	yes	---	yes	yes	yes	yes
Type of appointment	---	---	---	yes	yes	---	---	---	---	---
Number of months employed	yes	yes	yes	yes	yes	yes	yes	yes	yes	---
Percentage of time employed	yes	yes	yes	yes	yes	yes	yes	yes	yes	yes
Information on assignments	yes	---	yes	yes	---	yes	yes	yes	yes	yes

Teaching experience:										
Total years	yes	yes	yes	---	yes	yes	yes	yes	yes	yes
Years in current district	yes	---	yes	---	yes	yes	---	---	yes	---
Years in other district(s)	yes	---	---	---	yes	---	---	---	---	yes
Years in current assignment	---	---	---	---	---	---	---	---	---	---
For new hires:										
Occupation prior year	---	---	---	---	---	---	---	---	---	---
Location of occupation prior year	---	---	---	---	---	---	---	---	---	---
Other data:	---	---	---	z	---	aa	bb	---	---	dd
Courses/levels taught:	yes	---	yes	yes	---	yes	---	yes	yes	yes

Availability of State Data on Public School Professional Personnel--continued

	File characteristics by state			
	AS	GM	NM	PR
Is an automated file maintained?	yes	yes	yes	no
Earliest year available:	85	82	---	---
How often is the file updated? (daily, monthly, annually)	A	---	ee	---
Data elements in files				
School, personal employment data:				
School district code	---	---	yes	---
School code	---	yes	yes	---
Social Security number	yes	yes	yes	---
Date of birth/age	yes	yes	yes	---
Sex	yes	yes	yes	---

Racial/ethnic group	yes	yes	yes	---
Ed. level/degree status	yes	yes	yes	---
Acad. major (bachelor's)	yes	yes	yes	---
Annual contract salary	---	---	yes	---
Type of appointment	yes	yes	yes	---
Number of months employed	yes	---	yes	---
Percentage of time employed	---	---	yes	---
Information on assignments	---	---	yes	---

Availability of State Data on Public School Professional Personnel--continued

	File characteristics by state			
	AS	GM	NM	PR
Teaching experience:				
Total years	yes	---	yes	---
Years in current district	yes	yes	yes	---
Years in other district(s)	---	---	yes	---
Years in current assignment	yes	---	yes	---
For new hires:				
Occupation prior year	---	yes	yes	---
Location of occupation prior year	---	---	yes	---
Other data:	---	---	ff	---
Courses/levels taught:	---	---	---	---

AS - American Samoa
GM - Guam
NM - Northern Marianas
PR - Puerto Rico

a. Optional, at individual's discretion.
b. Types and subjects of certification.
c. Number of years in state public schools.
d. Also, administrative experience.
e. Bimonthly.
f. Monthly contract salary; monthly state minimum salary.
g. Certificate type.
h. Professional code.
i. Years of training; semester-hours training.
j. Years in private education; reason for leaving, last course data (in service, etc.).
k. Certification status/type/subject; state of residence; separation date/cause.
l. Name of college/university; type of certificate.
m. College code; source of funding; certification status.
n. Certification status.
o. Reason for leaving.
p. Teaching half/full day; certification status.
q. "Experience status"--notes whether first-year teacher, employed in system last year, or returned to teaching after absence from public education, etc.
r. Certification area(s); NTE score; college graduated from.

s. College graduated from and year; certification information.

t. Years in military.

u. Years in/outside Oregon; extra duty assignments; reasons for leaving district.

v. Certification areas.

w. Certification areas.

y. Age optional; certification status.

z. Latest 5-year employment history.

aa. Years taught in VA; type of certificate; courses endorsed to teach.

bb. Whether new, reentering or transferring; past employment experience.

cc. As needed, but at least every 5 years.

dd. Hours required to renew certificate; endorsement areas; validity period of certificate.

ee. Updated by semester.

ff. Certification information; other courses required, completed.

Appendix D
Acronymns

American Association of Colleges for Teacher Education (AACTE)
Center for Policy Research in Education (CPRE)
Education Commission of the States (ECS)
Educational Testing Service (ETS)
International Association for the Evaluation of Educational Achievement (IEA)
International Educational Assessment program (IEA)
Local education agency (LEA)
Massachusetts Institute for Social and Economic Research (MISER)
Mathematical Association of America (MAA)
Mathematical Sciences Education Board (MSEB) of the National Research Council
National Assessment of Educational Progress (NAEP)
National Association of State Directors of Teacher Education and Certification (NASDTEC)
National Center for Education Statistics (NCES)
National Council of Teachers of Mathematics (NCTM)
National Education Association (NEA)
National Education Longitudinal Study of 1988 (NELS:88)
National Governors' Association (NGA)
National Longitudinal Survey (NLS)
National Research Council (NRC)
National Science Foundation (NSF)
National Science Teachers Association (NSTA)
National Teacher Examination (NTE)
Northeast Teacher Supply and Demand (NETSAD) study

Recent College Graduates (RCG) Surveys
Schools and Staffing Survey (SASS)
State education agency (SEA)
Western Interstate Commission for Higher Education (WICHE)

Appendix E
Biographical Sketches

WILMER S. CODY is state superintendent of education for Louisiana, responsible for elementary and secondary education, postsecondary vocational-technical schools, and state schools for handicapped youth. Previous assignments include teacher, principal, and three local school system superintendencies: Chapel Hill, North Carolina; Birmingham, Alabama; and Montgomery County, Maryland. Additional assignments include the planning and establishment of the National Institute of Education and planning a state-by-state comparison of student achievement for the Council of Chief State School Officers. For four years he chaired the Assessment Policy Committee of the National Assessment of Educational Progress. He received an A.B. from Harvard College in social relations, an Ed.M. degree in teaching, and an Ed.D. degree in school administration from the Harvard Graduate School of Education.

GLENN A. CROSBY is professor of chemistry and chemical physics at Washington State University. A member of several national committees concerned with the status of education in the United States, he also serves as a consultant to the American Chemical Society Committee on Education. He has been honored both locally and nationally as a teacher and educator. He is a recipient of the faculty excellence award at his home institution and four national awards in chemical education. Internationally recognized for his research contributions in the field of photophysics and photochemistry of transition-metal complexes, he has lectured widely in the United States, Germany, New Zealand, Australia, and Japan. He received a B.S. in chemistry and mathematics from Waynesburg College and a Ph.D. from the University of Washington. He was a postdoctoral fellow at Florida State

University, a Fulbright fellow (1964) and a Humboldt awardee (1978-79) in West Germany.

F. JOE CROSSWHITE is professor of mathematics and education at Northern Arizona University and professor emeritus of mathematics education at the Ohio State University. He has served as president of the National Council of Teachers of Mathematics, as chairman of the Conference Board of the Mathematical Sciences, and as a member of the Mathematical Sciences Education Board. His principal fields of interest are mathematics teacher education and school mathematics curricula. He received a Ph.D. in mathematics education from the Ohio State University.

HARRIET FISHLOW is coordinator of undergraduate enrollment planning in the University of California's university-wide administration. She is the developer of the University of California's undergraduate enrollment potential projection model and serves on the Teacher Supply and Demand Steering Committee of the California Commission on Teacher Credentialing. She received a B.S. degree in education from the University of Pennsylvania and M.A. and Ph.D. degrees in demography from the University of California, Berkeley.

DOROTHY M. GILFORD served as study director of the panel's work. Formerly, she served as director of the National Center for Education Statistics and as director of the mathematical sciences division of the Office of Naval Research; currently she is director of the National Research Council's Board on International Comparative Studies in Education. Her interests are in reseach program administration, organization of statistical systems, education administration, education statistics, and human resource statistics. A fellow of the American Statistical Association, she has served as vice president of the association and chairman of its committee on fellows. She is a member of the International Statistics Institute. She received B.S. and M.S. degrees in mathematics from the University of Washington.

F. THOMAS JUSTER (Chair) is a research scientist at the Institute for Social Research and professor of economics at the University of Michigan. He is currently a senior adviser for the Brookings Panel on Economic Activity and chair of the American Economic Association Committee on the Quality of Economic Statistics. He is a fellow of the American Statistical Association. He received a B.S. degree from Rutgers University and a Ph.D. degree in economics from Columbia University.

CHARLOTTE V. KUH is executive director of the Graduate Record Examinations Program at the Educational Testing Service. A labor economist, she has held teaching positions at the Harvard Graduate School of Education and at Stanford University. She also spent eight years as a manager at AT&T. She received a B.A. from Radcliffe College and a Ph.D. in economics from Yale University. As a researcher, she specializes in the economics of higher education and on forecasting demand and supply for highly trained personnel, especially those in science and engineering. In her current position, she is interested in the appropriate use of standardized tests in graduate admissions, in ways to increase minority participation in graduate education, and in the challenges graduate education will face in the 1990s.

EUGENE P. MCLOONE is professor of education in the Department of Education Policy, Planning, and Administration of the College of Education at the University of Maryland. He also is an associate professor of economics at the university. His research interests are in school finance, teacher retirement systems, education statistics, and information systems for educational policy making. He directed a congressionally mandated study for the National Center for Education Statistics and was associate director of research for the National Education Association. He received a Ph.D. in educational administration from the University of Illinois, with subspecialties in public finance and mathematical statistics.

MICHAEL MCPHERSON is professor and chairman of the Economics Department at Williams College. He has served as senior fellow at the Brookings Institution and as a member of the Institute for Advanced Study. McPherson is coeditor of the journal *Economics and Philosophy* and a contributing editor of *Change* magazine. He writes on ethics and on the economics of higher education. He received A.B., A.M., and Ph.D. degrees from the University of Chicago.

RICHARD J. MURNANE is professor of education at the Graduate School of Education, Harvard University. He recently chaired the National Research Council's Committee on Indicators of Precollege Science and Mathematics Education and coedited (with Senta Raizen) the Committee's report entitled *Improving Indicators of the Quality of Science and Mathematics Education in Grades K-12.* His recent research has concerned the operation of teacher labor markets and the connections between education and the productivity of the work force.

INGRAM OLKIN is professor of statistics and education at Stanford University, where he has served on the faculty since 1961. He received undergraduate training at the City College of New York, a master's degree in mathematical statistics from Columbia University, and a doctorate degree in mathematical statistics from the University of North Carolina. He is a fellow of the Institute of Mathematical Statistics, the American Statistical Association, the Royal Statistical Society, and the International Statistical Institute. His research interests are in multivariate analysis and models in the behavioral, social, and educational sciences. He has served as president of the Institute of Mathematical Statistics, has served on editorial boards for numerous journals, and has served on many government panels.

JOHN J. STIGLMEIER is director of the Information Center on Education in the New York State Education Department. In that position, he is responsible for the development of educational information systems as well as the analysis, interpretation, and dissemination of data relating to the state's educational enterprise. He participates actively in the Council of Chief State School Officers' Committee on Evaluation and Information Systems. Recently, he conducted a month-long seminar on educational statistics in the People's Republic of China. He received a B.S. in biology from Siena College, an M.S. in reading education from the State University of New York, Albany, and a Ph.D. in educational psychology from Forham University.

ELLEN TENENBAUM, a public policy and survey research analyst, served as consultant to the panel. Her work at the National Research Council since 1981 has spanned studies concerning education and minerals statistics and energy and natural resources policy. She has a master's degree in public policy from the University of California at Berkeley.